Synthesis Lectures on Engineering, Science, and Technology

The focus of this series is general topics, and applications about, and for, engineers and scientists on a wide array of applications, methods and advances. Most titles cover subjects such as professional development, education, and study skills, as well as basic introductory undergraduate material and other topics appropriate for a broader and less technical audience.

Khaled Salah Mohamed

Next Generation EDA Flow

Motivations, Opportunities, Challenges
and Future Directions

 Springer

Khaled Salah Mohamed
Siemens
Fremont, CA, USA

ISSN 2690-0300   ISSN 2690-0327 (electronic)
Synthesis Lectures on Engineering, Science, and Technology
ISBN 978-3-031-88434-4   ISBN 978-3-031-88435-1 (eBook)
https://doi.org/10.1007/978-3-031-88435-1

This Springer imprint is published by the registered company Springer Nature Switzerland AG
The registered company address is: Gewerbestrasse 11, 6330 Cham, Switzerland

If disposing of this product, please recycle the paper.

Dedicated to my Brave Sister Dr. Dina Salah,

In the face of numerous challenges and daunting situations, you have stood resolute, embodying strength and courage with unwavering resolve. Your journey has been a testament to your unwavering spirit and indomitable will. Through every trial, you have shown remarkable resilience, facing adversity with grace and determination. Your bravery knows no bounds, and your unwavering determination inspires us all. As you continue to tread the path ahead, I wish you boundless success and triumph in all your endeavors. May you continue to persevere, thrive, and emerge victorious in every facet of life.

With admiration and love,

Preface

In today's rapidly evolving landscape of electronic design, the emergence of new and unconventional technologies presents both exciting opportunities and unique challenges. These innovations may seamlessly integrate with existing Electronic Design Automation (EDA) tools, or they may necessitate the development of novel EDA tools tailored to their specific requirements. Thus, understanding the anatomy of EDA tools becomes paramount for engineers.

This book embarks on a journey to explore the inner workings of EDA tools, offering a comprehensive and insightful examination of their functionality and capabilities. Through thoughtful analysis and practical examples, we delve into the core principles behind EDA tools, shedding light on their role in shaping the future of electronic design.

With a meticulous focus on numerical methods, this book delves deep into the mathematical foundations that underpin EDA tools. From finite element analysis to Monte Carlo simulations, readers will gain a thorough understanding of the numerical techniques employed to model and simulate complex electronic systems.

Furthermore, this book elucidates the diverse modeling methods utilized in EDA tools, providing readers with a holistic view of the strategies employed to represent and analyze electronic circuits and systems. Whether exploring circuit-level simulations or system-level modeling, this text equips readers with the knowledge needed to navigate the intricacies of EDA toolsets.

Moreover, as we stand at the threshold between quantum and classical computing paradigms, this book delves into the fascinating intersection of quantum mechanics and electronic design. By examining the evolving landscape of quantum EDA tools, readers will gain insights into the transformative potential of quantum computing in electronic design.

Lastly, this book explores the transformative impact of machine learning on EDA tools, offering insights into how artificial intelligence techniques can enhance performance and productivity. From automated design optimization to predictive modeling, machine learning algorithms are poised to revolutionize the EDA landscape, enabling engineers to tackle increasingly complex design challenges with unprecedented efficiency.

This book serves as a comprehensive guide to the world of EDA tools, offering readers a deeper understanding of their inner workings and a glimpse into the future of electronic design. Whether you are a seasoned engineer, a researcher, or a student embarking on a journey in electronic design, this book is designed to be your companion in unlocking the full potential of EDA tools.

Overview

- New unconventional technologies may use the existing EDA tools or they may need new simulators. Thus, it is important to understand how EDA tools work. This book thoughtfully discusses how EDA tools work.
- This book provides a deep dive into different numerical methods used in EDA tools.
- This book addresses the different EDA tools in modeling methods.
- This book discusses the threshold between quantum and classical EDA tools.
- This book discusses how machine learning can be deployed in EDA tools to enhance performance and productivity.

Fremont, USA   Khaled Salah Mohamed

Competing Interests The author has no competing interests to declare that are relevant to the content of this manuscript.

Contents

The Next Wave of Moore's Law

1

"Technology Level, Architectural Level, Circuit/Logic Level, and Software/OS Level"

1.1 Introduction

Moore's Law is an observation that the number of transistors on a microchip doubles every two years. Continuous advancement in semiconductor manufacturing processes, leading to the production of smaller, faster, and more efficient integrated circuits. Over the years, the semiconductor industry has witnessed a steady progression in technology nodes, with each new node representing a reduction in transistor size and an increase in transistor density. This evolution has been driven by the pursuit of Moore's Law, which predicts that the number of transistors on a chip will double approximately every two years. As technology nodes advance (Table 1.1), semiconductor manufacturers adopt new materials, equipment, and fabrication techniques to achieve smaller feature sizes and improve chip performance. This evolution has enabled the development of increasingly powerful and energy-efficient electronic devices, from smartphones and laptops to data centers and automotive systems. Key milestones in technology node evolution include the transition from planar to FinFET transistors, the introduction of advanced lithography techniques such as EUV (Extreme Ultraviolet), and the integration of new materials like high-k dielectrics and metal gates. However, nowadays Interconnect dimensions and CMOS transistor feature sizes approach their physical limits. Therefore, scaling will no longer be the sole contributor to performance improvement. So, instead of trying to improve the performance of traditional CMOS circuits, integration of multiple technologies and different components in a heterogeneous system that is high performance will be introduced "More than Moore" and CMOS replacement" beyond CMOS" will be explored. The next wave of Moore's Law is expected to come from new technologies that enable more efficient use of existing transistors, as well as new materials and structures that allow for the creation of smaller, more powerful transistors. This chapter focuses on technology level trends, where it presents "More Moore": new architectures (SOI, FinFET, Twin-Well)," More Moore":

© The Author(s), under exclusive license to Springer Nature Switzerland AG 2025
K. S. Mohamed, *Next Generation EDA Flow*, Synthesis Lectures on Engineering, Science, and Technology, https://doi.org/10.1007/978-3-031-88435-1_1

Table 1.1 Technology node evolution

Technology node (nm)	Feature size (nm)	Year	Transistor technology
180	180	1999	Planar CMOS
130	130	2001	Planar CMOS
90	90	2004	Planar CMOS
65	65	2006	Planar CMOS
45	45	2008	Planar CMOS
32	32	2010	Planar CMOS
22	22	2012	FinFET
14	14	2014	FinFET
10	10	2017	FinFET
7	7	2018	FinFET
5	5	2020	FinFET
3	3	2022	FinFET
2	2	TBD	FinFET

new materials (High-K, metal gate, strained-Si)," More than Moore": new interconnect schemes (3D, NoC, optical, wireless), and "Beyond CMOS": new devices (molecular computer, biological computer, quantum computer). Three-dimensional integration is a promising alternative option to traditional 2D planar chips. 3D integration is mainly concerned with the communication infrastructure between different stacked dies of future multi-core SoC and network-on-chip (NoC). Among several 3D integration technologies, the TSV (through silicon via) approach is the most promising one and therefore is the focus of the majority of 3D integration R&D activities. However, there are challenges that should be overcome before the production of TSV-based 3D ICs becomes possible, e.g., electrical modeling challenges, thermal and power challenges, technological challenges, design methodology challenges, and CAD tool development challenges. The manufacturability of TSV-based 3D-ICs is an important issue for realizing real 3D-ICs designs. It is mandatory to simulate and verify everything. Moreover, new unconventional technologies may use the existing EDA tools or need new EDA tools. So, it is important to understand how EDA tools work [1].

1.2   Work Around Moore's Law

Clearly, devices cannot be fabricated smaller than the dimension of a single molecule. Also, lithography technology seems unfit for atomically precise manufacturing and if silicon dioxide insulators are reduced to just a few atomic layers, electrons can tunnel directly through the gate. The interconnect congestion bottleneck is also a limiting factor. These limitations of silicon-based ICs are summarized in Table 1.1. These limitations are now causing the industry to identify at least three main research domains at technology levels (Fig. 1.1), called: (1) "More Moore", (2) "More than Moore", and (3) "Beyond CMOS". The "More Moore" domain traditionally deals with technologies related to the silicon-based CMOS. The "More than Moore" domain encompasses the engineering of complex systems that can combine, by heterogeneous integration techniques (in SoC or SIP), various technologies (not exclusively electronic) to meet certain needs and challenging specifications of advanced applications. The "Beyond CMOS" domain deals with new technologies and device principles (i.e., from charge-based to non-charge-based devices, from semiconductor to molecular technology) [2–4]. Table 1.2 shows different trends to overcome Moore's law saturation at different levels: technology level, architectural level, circuit/logic level, software/OS level, and packaging level. All levels are working simultaneously to improve the performance (Table 1.3).

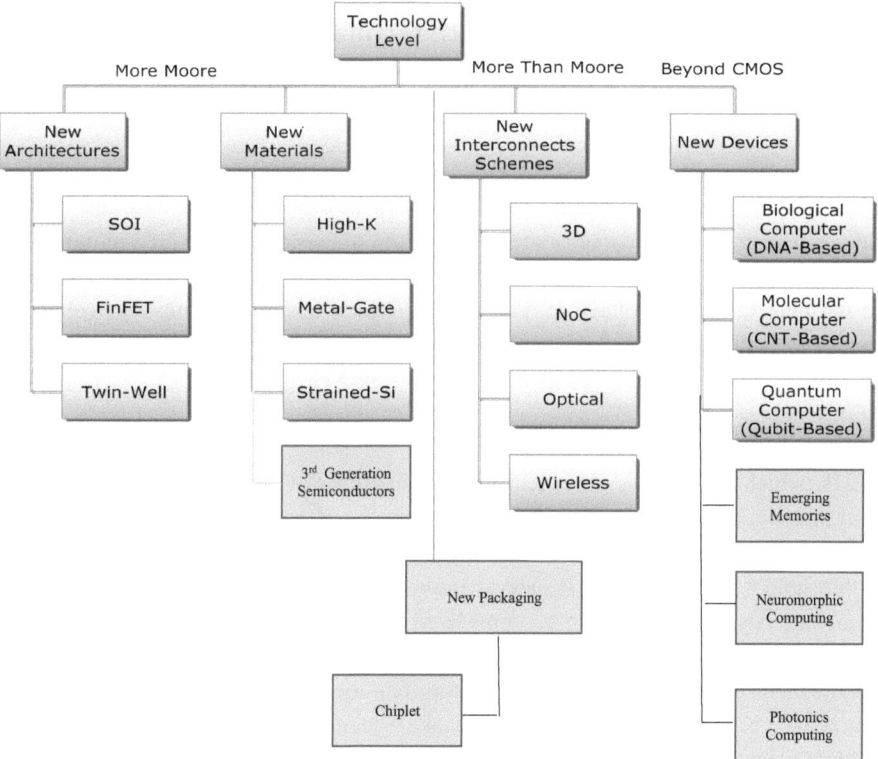

Fig. 1.1 Summary of solutions on the technology level to work around Moore's law. It includes incremental changes and fundamental shift

Table 1.2 **Transistor scaling**: physical limitations of silicon-based ICs

Manufacturing limitations	
Lithography	Lithography technology seems unfit for atomically precise manufacturing
Transistor dimensions	Transistor dimensions are approaching a hard limit that cannot be overcome. That limit is the size of the atom and molecule. Clearly, devices cannot be fabricated smaller than the dimension of a single molecule [5]
Material limitations	
SiO_2	If silicon dioxide insulators are reduced to just a few atomic layers, electrons can tunnel directly through the gate
Design limitations	
Interconnects bottleneck	Interconnect delay becomes dominant over gate delay

1.3 Work Around Moore's Law: Technology-Level Trends

1.3.1 More Moore (MM): New Architectures

a. **SOI**

The basic concept of silicon on insulator is simple. Rather than fabricating a transistor whose body is connected to the substrate, which is the normal method, an insulating oxide is deposited on the Si-substrate and then the transistor is fabricated on top of that. By doing this, the body is then electrically isolated from its surroundings. This means that the bulk to source voltage is now floating which lowers the threshold voltage and capacitance, providing a performance increase [5].

b. **FinFET**

FinFET, which is a double-gate field effect transistor (DGFET), is more versatile than traditional single-gate field effect transistors because it has two gates that can be controlled independently. Usually, the second gate of FinFETs is used to dynamically control the threshold voltage of the first gate to improve circuit performance and **reduce leakage power**. Also, a FinFET can be considered a three-dimensional version of MOSFET (Fig. 1.2) [6].

Nanosheet FETs are advanced transistors featuring a gate-all-around (GAA) architecture, where the gate surrounds the channel on all sides. This design significantly enhances electrostatic control over the channel and reduces short-channel effects, leading to improved performance and lower leakage compared to finFETs. The key advantage of nanosheet FETs is their ability to deliver higher drive current by increasing channel width

Table 1.3 Comparison between different trends levels from performance gains point of view

Technology level	Trends	Performance gains					
		Area	Speed throughput delay	Power	Noise	Thermal	Yield reliability
More Moore	Scaling	✓	✓	✓			
	New architectures						
	SOI (silicon on insulator)	✓					
	Twin-Well			✓			
	Fin-FET		✓				
	New materials						
	High-K		✓				
	Metal-G		✓				
	Strained-Si		✓				
More than Moore	*New interconnects schemes*						
	NoC (network on chip)		✓	✓			
	3D (three dimensional)		✓	✓		x	x
	Optical-interconnects		✓	✓		x	
	Wireless interconnects	✓			x		
	New packaging	✓					
Beyond CMOS	*New devices*						
	Molecular computer CNT (carbon nano tubes) (nano wires) (quantum-Dot)	✓	✓				
	Biological computer (DNA-Computing)		✓				x
	Quantum computer (SET: single electron transistor) (spin device)	✓					

(continued)

Table 1.3 (continued)

Trends		Performance gains						
		Area	Speed throughput delay	Power	Noise	Thermal	Yield reliability
Architectural level	Emerging memories	✓	✓				
	Neuromorphic computing		✓				
	Mc (multi-core)	✓	✓				
	DM (distributed memory)	✓					
	HW/SW co-design		✓				
	In-memory computing		✓				
Circuit/logic level	Adiabatic logic		✓				
	MTCMOS		✓				
	Multiple-Vdd		✓				
	Clock–gating		✓				
	Power–gating		✓				
	Asynchronous circuits		✓				
	Pipelining		✓				
	Data–encoding	✓	✓				
	Repeater	✓	✓				
Software/OS level	Concurrency		✓				
	Partitioning		✓				
	Sleep mode		✓				
Packaging level	Heat sink				✓		

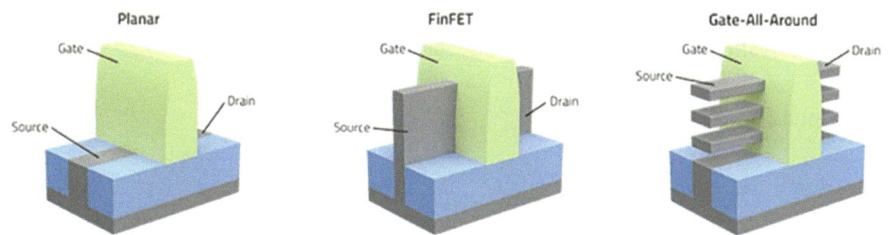

Fig. 1.2 Planar transistors versus FinFETs versus gate-all-around source

within the same circuit footprint. This is achieved by stacking multiple nanosheet channels vertically, which boosts current capabilities without expanding the device area. The
fabrication process of nanosheet FETs starts with the deposition of a silicon/silicon–germanium (Si/SiGe) heterostructure, which is isolated from the substrate to prevent parasitic
conduction. The heterostructure is then patterned into pillars through anisotropic etching,
forming the foundation of the nanosheet channels. An inner spacer etch step, following
the creation of a dummy gate, is crucial as it defines the gate length and source/drain
junction overlaps by creating a recess in the SiGe layers. This process is completed with
source/drain epitaxy, a channel release etch, and the formation of the replacement gate.
Nanosheet FETs also benefit from the flexibility provided by extreme ultraviolet lithography, allowing for variable device widths and more efficient use of space. The transition
from silicon orientation (110) to (100) further modifies carrier mobilities in the channel,
potentially enhancing overall transistor performance.

c. **Twin-Well**

For high-performance chips, a low doped substrate is used, and then constructs two wells
are constructed at optimum doping levels. Since the substrate is lightly doped, there is
less chance for latch-up because of the high resistivity [6].

1.3.2 More Moore: New Materials

a. **High-K Dielectric**

High-K dielectrics are designed to address one aspect of off-state power consumptions:
gate tunneling currents, as it reduces gate leakage [4].

b. **Metal-Gate**

The doped polycrystalline silicon used for gates has a very thin depletion layer, approximately 1 nm thick, which causes scaling problems for small devices. Other metals are

being investigated for replacing the silicon gates, including tungsten and molybdenum to eliminate polysilicon depletion and for thermal stability [4].

c. Strained-Si

If we cannot make MOSFETs yet smaller, instead move the electrons faster. Strained Si is the process of introducing physical strain on the silicon lattice to help improve electron and hole mobility. This allows the holes and electrons to flow more freely, reducing device resistance and other properties affected by electron/hole mobility [5].

d. Third-Generation Semiconductors

During the development of semiconductor industry, silicon (Si), and germanium (Ge) are called the first generation of semiconductors. Gallium arsenide (GaAs) and (GaSi) represent the second generation of semiconductors. The third-generation semiconductors have materials like gallium nitride (**GaN**) and silicon carbide (**SiC**), a must-have technology for many manufacturers at this point of time. They are suitable for high temperature, high frequency, radiation resistant and high power devices like electric vehicles, data centers, and renewable energy production. While SiC has higher electron mobility than Si, GaN's electron mobility is higher than SiC meaning that GaN should ultimately be the best device for very high frequencies. Higher thermal conductivity means that the material is superior in conducting heat more efficiently [7].

1.3.3 More Than Moore (MTM): New Interconnects Schemes

a. 3D Integration

Increasing drive for the integration of disparate signals (digital, analog, RF) and technologies (SOI, SiGe, HBTs, GaAs, strained silicon, and so on) results in introducing various design concepts, for which existing planar (2D) technologies may not be suitable, so, three-dimensional (3D) integrated circuits are introduced. 3D integration technology provides increased performance in many design criteria as compared to the current 2D approaches. 3D-ICs, which contain multiple layers of active devices, extensively utilize the vertical dimension to connect components and are expected to address interconnect delay related problems in planar (2D) technologies, by the use of short wires in 3D designs. These shorter wires will decrease the average load capacitance and resistance and decrease the number of repeaters which are needed to regenerate signals on long wires and to enable the integration of heterogeneous technologies. In the 3D design, an entire (2D) chip is divided into a number of different blocks, and each one is placed on a separate layer of silicon that are stacked on top of each other. This may be exploited to

Fig. 1.3 3D integration
example of two dies

build SoC by placing different circuits with performance requirements in different layers
(Fig. 1.3) [8–10].

b. **NoC**

Network on chip (NoC) is a promising solution to simplify and optimize SoC design.
NoCs consist of switches and links and use circuit or packet switching to transfer data
through the system. NoC have been proposed as a shared communication medium that is
highly scalable and can offer enough bandwidth to replace many traditional bus-based and
point-to-point links [11]. A Network on Chip (NoC) is a communication architecture used
in computer systems, particularly in digital circuits and systems-on-chip (SoCs). It is a
distributed communication system that replaces traditional point-to-point communication
between different components of a computer system with a network-like infrastructure.
In an NoC, the components of a computer system (e.g., processors, memory, input/output
interfaces, etc.) are connected through a network of interconnects, similar to the way
computers are connected in a local area network (LAN). Each component is connected
to the network through a network interface, and communication between components
is accomplished by sending packets of data over the network. NoCs offer several advantages over traditional point-to-point communication architectures. They can scale to larger
systems with many components more easily, as adding new components simply requires
connecting them to the network rather than adding more point-to-point connections. NoCs

also typically provide more efficient communication between components, as packets can be routed dynamically through the network to avoid congestion and optimize performance. NoCs are commonly used in modern digital circuits and SoCs, particularly in high-performance computing applications. They can be implemented in hardware or in software, depending on the specific requirements of the system. Specialized design tools and languages, such as SystemC and SystemVerilog, are often used to design and simulate NoCs [12].

c. **Optical Interconnects**

Thanks to the unique properties of optical communication, such as bit-rate transparency and low loss optical waveguides, photonic NoCs have been proposed as an optimal solution to reduce the overall power budget [6]. Optical interconnects are a type of communication technology that uses light instead of electrical signals to transmit data between different components in a computer system. Optical interconnects are often used to connect different components within a single system or between systems, and they offer several advantages over traditional electrical interconnects. One of the main advantages of optical interconnects is their high bandwidth. Optical interconnects can transmit data at much higher speeds than electrical interconnects, which is particularly important in high-performance computing applications. Optical interconnects can also transmit data over longer distances than electrical interconnects, which makes them useful in applications where components are physically separated. Another advantage of optical interconnects is their low power consumption. Transmitting data using light requires much less power than transmitting data using electrical signals, which can be a significant advantage in low-power applications or in large-scale systems where power consumption is a concern. Optical interconnects can be implemented using a variety of different technologies, including fiber optics, integrated optics, and free-space optics. Fiber optics use thin strands of glass or plastic to transmit light, while integrated optics use optical components that are integrated onto a single chip. Free-space optics use beams of light that are transmitted through the air or through a vacuum. Optical interconnects are commonly used in modern computer systems, particularly in high-performance computing applications such as data centers and supercomputers. They are also used in some consumer devices, such as high-end gaming computers and virtual reality headsets.

d. **Wireless Interconnects**

Unlike 3D and photonic NoCs, NoC with RF interconnects can be built using existing CMOS technology [13]. The process of building an NoC with RF interconnects typically involves integrating a RF front-end circuitry into the NoC design. The RF front-end circuitry consists of components such as antennas, amplifiers, filters, and mixers that are used to transmit and receive wireless signals. To integrate the RF front-end circuitry into the NoC design, the existing CMOS fabrication processes can be modified to include additional layers of metal interconnects, which are used to create the RF circuitry. The RF circuitry can then be integrated with the digital circuitry of the NoC using standard CMOS design methodologies, such as HDL (Hardware Description Language) and EDA (Electronic Design Automation) tools. One of the key advantages of using existing CMOS technology to build an NoC with RF interconnects is that it leverages the economies of scale that have been developed for CMOS technology over the years. This can help to reduce the cost and time required to develop and produce an NoC with RF interconnects, making it more accessible to a wider range of applications.

1.3.4 More Than Moore (MTM): New Packaging Technologies

a. **Chiplet**

Chiplets are a key component of the "Beyond Moore's Law" paradigm, which refers to the shift in the semiconductor industry towards new design methodologies and packaging technologies to enable continued performance improvements beyond the limitations of traditional scaling of integrated circuits. Chiplets are individual semiconductor chips that can be combined to form a larger system-on-chip (SoC), enabling greater flexibility and scalability in chip design. Chiplets can be optimized for specific functions, such as graphics processing, memory, or data transfer, and then combined with other chiplets to form a complete SoC. This approach allows for greater modularity in chip design, making it easier to design and manufacture complex systems with a high degree of customization and specialization. Chiplets are relevant at multiple design levels, from the transistor level to the system level. At the transistor level, chiplets can be used to implement specialized functions, such as analog circuits, that are difficult to integrate into traditional monolithic designs. At the logic level, chiplets can be used to implement specific functions, such as memory or processing units, that can be combined in different configurations to meet the requirements of different applications. At the system level, chiplets can be combined to form a complete SoC, enabling greater flexibility in system design and allowing for customization to specific use cases [14]. Chiplet is a solution for many SoC limitations. Chiplets enhance chip yields and reduce costs while maintaining the performance level of a single, large monolithic chip. Designers have the flexibility to combine various chiplets, leverage optimal process technologies for specific functions, utilize Chiplet intellectual property (IP), streamline transitions to new process nodes, and mitigate issues related to wafer waste and manufacturing defects. Chiplets play a pivotal role in enabling the creation of exceptionally dense and high-performance chips essential for contemporary networking, storage, artificial intelligence/machine learning (AI/ML), analytics, media processing, high-performance computing (HPC), and virtual reality applications.

b. Embedded Multi-die Interconnect Bridge (EMIB)

Embedded Multi-die Interconnect Bridge (EMIB) technology is a packaging technology used for chiplets, which are individual semiconductor chips that can be combined to form a larger system-on-chip (SoC). EMIB is a method of connecting multiple chiplets together using a small, high-bandwidth interconnect bridge that is embedded in the package. The EMIB technology involves embedding a silicon bridge on the package substrate, which is used to connect the individual chiplets. The bridge provides a high-speed and low-latency interconnect that enables the chiplets to communicate with each other as if they were on a single chip. This technology allows for the creation of highly integrated and customized systems, by combining different chiplets that are optimized for specific functions, such as graphics processing, memory, or data transfer. EMIB technology has several advantages over traditional packaging methods, such as wire bonding or flip-chip interconnects. For example, EMIB allows for higher bandwidth and lower power consumption, as well as improved thermal management. Additionally, EMIB enables greater flexibility and scalability in chip design, as chiplets can be added or removed as needed to meet the specific requirements of a given application.

c. Multi-Chip Module (MCM)

MCM is a packaging technology that combines multiple chips, or dies, into a single module. The individual chips are mounted on a substrate and connected with wires or conductive pathways to create a larger system. MCM can contain a mix of different types of chips, such as microprocessors, memory chips, and specialized logic circuits, which are optimized for specific functions.

d. Monolithic Die

It refers to a single semiconductor chip that contains all the necessary components, such as logic gates, memory cells, and input/output interfaces, to perform a specific function. The entire chip is fabricated on a single substrate using traditional semiconductor manufacturing techniques.

e. Silicon Interposer

It is a packaging technology that uses a silicon substrate to interconnect multiple semiconductors dies. The interposer contains a network of metal traces that provide high-speed connections between the dies, as well as through-silicon vias (TSVs) that provide vertical connections. Silicon Interposer technology can provide a high degree of design flexibility, enabling a wide range of chip-to-chip connections and configurations [1, 15]. Embedded Multi-Die Interconnect Bridge (EMIB) versus Silicon interposer versus Multi-Chip Module (MCM) Package versus Monolithic Die is shown in Fig. 1.4 and Table 1.4.

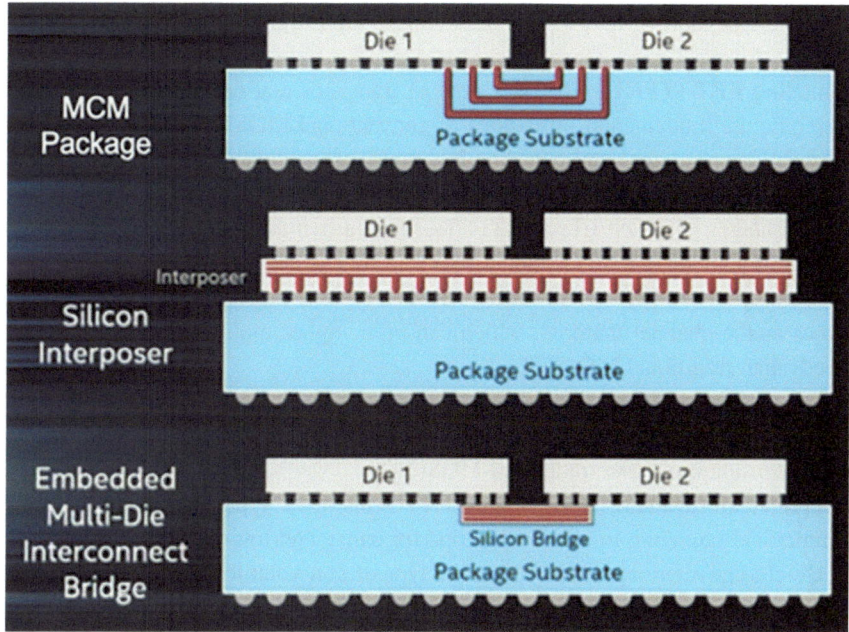

Fig. 1.4   EMIB versus silicon interposer versus MCM Package

Table 1.4   Comparison between EMIB, silicon interposer, MCM Package and Monolithic Die

Technology	Advantages	Disadvantages	Relevant technology theory
EMIB	High-speed interconnects, low power consumption	Limited design flexibility, only for Intel processors	3D integration, fine-pitch interconnects
Silicon interposer	High degree of design flexibility, high-speed interconnects	Higher cost, lower yield, requires specialized manufacturing	Through-silicon vias, 2.5D/3D integration
MCM Package	High performance, low power consumption	Higher cost, more complex design and testing	Die stacking, wire bonding, flip-chip bonding
Monolithic Die	Simple and cost-effective to manufacture	Limited performance and functionality, not scalable	CMOS process, planar technology

1.3.5   Beyond CMOS: New Devices

Beyond-CMOS device technologies refer to non-conventional device technologies that are being researched and developed as alternatives to traditional complementary metal–oxide–semiconductor (CMOS) technology, which is widely used in today's digital circuits. Some of the notable beyond-CMOS device technologies are [16]:

- **Tunneling FET (TFET):** This is a type of transistor that uses quantum tunneling to overcome the fundamental limitations of conventional MOSFETs. TFETs can operate at much lower voltages and consume less power than traditional MOSFETs.
- **Hybrid phase transition FET (HyperFET):** This is a type of transistor that combines the advantages of MOSFETs and TFETs, using a hybrid structure that leverages both quantum tunneling and conventional transport mechanisms.
- **Carbon nanotube FET (CNTFET):** This is a type of transistor that uses carbon nanotubes as the channel material, which can offer higher mobility and faster switching speeds than traditional MOSFETs.
- **Silicon nanowire FET (SiNWFET):** This is a type of transistor that uses silicon nanowires as the channel material, which can offer higher performance and lower power consumption than traditional MOSFETs.
- **Symmetrical tunneling FET (SymFET):** This is a type of transistor that uses symmetrical tunneling to achieve high-performance and low-power operation.
- **Phase-change memory (PCM):** This is a type of non-volatile memory that uses phase-change materials to store data. PCM can offer higher density and faster access times than traditional flash memory.
- **Spin-transfer torque magnetic tunnel junction (STT-MTJ):** This is a type of memory device that uses magnetic tunnel junctions to store data. STT-MTJs can offer higher density and faster access times than traditional magnetic memory.
- **Resistive random access memory (RRAM):** This is a type of non-volatile memory that uses the resistance of a thin film to store data. RRAM can offer higher density and faster access times than traditional flash memory.

Moreover, we can categorize Beyond CMOS devices inn three categories described in the below sub-sections.

a. Biological Computer: DNA-Based

DNA has the two properties that are required to do computing: a way to store information and a means of manipulating information. Any system having these two necessary properties can be setup to do that computation. Cells (living parts) of organisms are ingredients for computation. These provide the basic idea of computing, as these tiny parts are complete machines and perform all the processing for the organisms' activities [17]. It can overcome on two major limitations of silicon-based traditional computers: storage capacity and processing speed. In this technique, enzymes and amino acids of DNA are used to solve problem. In this approach of computing some specific information is encoded on DNA and is then used to perform bimolecular processes to attain intended computing [18].

b. Molecular Computer: CNT-Based

Carbon Nanotube (CNT) is a new molecular structure of carbon and has been applied to semiconductor industry recently. It has unique electrical and mechanical properties that it can be shaped to act as conductor, semiconductor, and insulator. CNT can reduce semiconductor manufacturing processing steps that incorporate copper, such as the forming of insulator to prevent copper from diffusing through the gate. CNTs are cylindrical structures based on the hexagonal lattice of carbon atoms that covalently bond together and forms crystalline graphite. CNTs can be looked at as single molecules, due to their small size (~ 1 nm in diameter and ~ 1 μm length). There are two major types of CNTs: single-walled carbon nanotubes (SWNT) and multi-walled carbon nanotubes (MWNT) [18].

c. Quantum Computer/Ising Computing: Qubits-Based Computing

This concept is built around the theory that quantum particles can exist in many 'universes' at once and only collapse down to one 'universe' when observed. This could achieve massive parallelism in computation [5]. Silicon-based computers used transistors to build logic gates, but quantum computers use special quantum bits called qubits. Moreover, a group of qubits is called qubit register. Traditional computer bit has only two states either '0' or '1' but on the other hand a quantum computer can carry multiple values at the same time. Qubit can be either '0' or '1' and sometimes both '0', '1' at the same time and it is known as superposition of states. Quantum computation can be implemented for simplifying a variety of problems, such as search algorithms, cryptography number theory. Nevertheless, quantum computers cannot solve many other problems such as sorting data [18–20].

d. Emerging Memory and Storage Devices

SRAM/DRAM is fast but has large leakage power and volatile. Floating-gate based Flash is non-volatile but exhibits low writes speed and limited write endurance. Therefore, recent research focuses on hybrid memory structures to get the advantages of both. From the perspective of system level, 3D integration can be employed to integrate hybrid memory components with high density, where it can also reduce the distance between

Table 1.5 Emerging memories

CBRAM	Conductive bridge
ReRAM	Resistive
PCRAM	Phase change
FeRAM	Ferroelectric
ST-MRAM	Spin-torque magnet
Memristor	It is called the fourth element (change of flux with charge)

components to few micrometers instead of few centimeters. Emerging memory technologies are making steady progress towards product introductions, including phase-change memory (PCRAM), resistive memory (ReRAM), and magnetic memory (MRAM). The new trends in memories are summarized in Table 1.5. They provide higher density, lower latency, lower power per bit for both read and write operation, and high read/write/erase processing speed [21].

Resistive Random Access Memory (ReRAM)

Resistive Random Access Memory (ReRAM) stands out as a promising non-volatile memory technology characterized by its resistive switching behavior. This innovative memory architecture consists of two metal electrodes separated by a thin insulating layer, forming a Metal–Insulator–Metal (MIM) configuration crucial for enabling resistive switching. This phenomenon involves a sudden change in resistance when a strong electric field or current is applied to the dielectric layer. ReRAM operates based on two main types of resistive switching behavior: unipolar and bipolar, distinguished by their current–voltage (I-V) characteristics. The process begins with a high voltage triggering the "forming process," which transitions a memory cell from a highly resistive state to a low resistive state. Subsequently, the "reset process" reverses this transition by changing the low resistive state cell back to a highly resistive state via a threshold voltage (Vth). Conversely, the "set process" converts a highly resistive state to a low resistive state by applying a voltage greater than the reset voltage. This intricate switching behavior is observed in various highly insulating oxides, including binary metal oxides. ReRAM utilizes novel materials such as transition metal oxides and organic compounds strategically arranged to exhibit reversible resistive switching under voltage pulses. Examples of these transition metal oxides include TiO_2, NiO, $SrTiO_3$, Cr-doped $SrZrO_3$, Cu_2O, and PCMO [22].

e. **Neuromorphic Computing**

Neuromorphic computing aims to mimic the structure and function of the human brain in digital or analog hardware. Replicating the brain's capabilities in artificial neuromorphic computing has been challenging. The digital components of neuromorphic computing typically involve specialized hardware and algorithms designed to emulate the behavior of biological neural networks. Here are some key digital components of neuromorphic computing:

- **Spiking Neurons**
  - In neuromorphic computing, digital representations of spiking neurons are used to model the behavior of biological neurons. These neurons generate spikes or action potentials in response to input stimuli.
- **Synapses**
  - Digital synapses emulate the connections between neurons in the brain. These synapses play a crucial role in transmitting signals (spikes) between neurons. Weight values associated with synapses represent the strength of connections.
- **Neural Networks**
  - Digital neural networks in neuromorphic computing consist of interconnected spiking neurons and synapses. These networks can be organized in layers (e.g., input layer, hidden layers, output layer) and are trained to perform specific tasks using learning algorithms.
- **Learning Algorithms**
  - Neuromorphic systems often incorporate digital learning algorithms inspired by biological learning mechanisms. These may include supervised learning, unsupervised learning, and reinforcement learning algorithms.
- **Event-Driven Processing**
  - Unlike traditional von Neumann architectures (Fig. 1.5), neuromorphic computing often operates in an event-driven manner. Processing occurs only when relevant events, such as spikes, occur, reducing power consumption and mimicking the asynchronous nature of neural communication.
- **Memory Elements**
  - Memory elements store information related to the state of neurons and synapses. Digital representations of short-term and long-term memory are essential for learning and memory functions in neuromorphic systems.

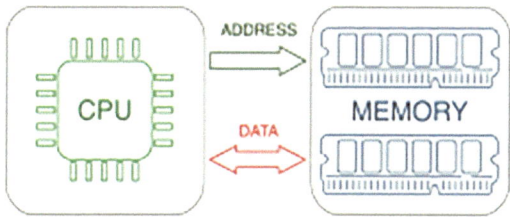

Fig. 1.5 Von-Neumann architecture

- **Parallel Processing**
  - Neuromorphic systems leverage parallel processing to handle multiple computations simultaneously, aligning with the parallel nature of information processing in biological brains.
- **Neuromorphic Chips or Processors**
  - Specialized digital hardware, often in the form of neuromorphic chips or processors, is designed to efficiently implement the computations involved in neuromorphic computing. These chips may have unique architectures optimized for spiking neural networks.
- **Software Frameworks**

  1. Digital components of neuromorphic computing are often supported by software frameworks that facilitate the design, simulation, and programming of neuromorphic systems. Examples include NEST, BindsNET, and Brian.

- **Neuromorphic Vision Systems**

  1. Some neuromorphic computing applications focus on mimicking the visual processing capabilities of the human brain. Digital components in these systems include algorithms for event-based vision processing.

Neuromorphic computing can be implemented using a variety of technologies, and it's not strictly limited to CMOS (Complementary Metal–Oxide–Semiconductor) technology. While traditional computing systems, including many digital components, are based on CMOS technology, neuromorphic computing explores different materials and architectures to better mimic the functioning of biological neural networks. In the context of neuromorphic computing, researchers and engineers may explore various technologies beyond CMOS to implement digital components. Some of these technologies include:

- **Memristors**: Memristors are resistive devices that can store and process information. They are considered promising for neuromorphic computing due to their non-volatile nature and potential for emulating synapses.
- **Spintronics**: Spin-based electronics use the spin of electrons to encode information. Spintronics can be explored in neuromorphic computing for efficient information processing.
- **Quantum Dots**: Quantum dots, semiconductor nanoparticles, can be utilized for implementing certain neuromorphic functionalities, especially in the context of quantum neuromorphic computing.
- **Photonic Neuromorphic Computing**: Optical or photonic approaches are explored for neuromorphic computing, leveraging light for information processing instead of traditional electronic signals [23].

f. Photonics Computing

Silicon photonics is a technology that integrates optical components, such as lasers, modulators, detectors, and waveguides, into the traditional semiconductor material, silicon. This integration allows for the generation, manipulation, and detection of light signals on a silicon chip, enabling the transmission of data using photons instead of electrons. Photonic computing stands at the forefront of promising "beyond-CMOS" technologies, charting a transformative course for the future of information processing. By harnessing the unique properties of light, photonic computing surpasses the constraints of traditional complementary metal–oxide–semiconductor (CMOS) technology. The use of photons for computation and communication not only enables blazingly fast data transmission but also holds the potential to address the burgeoning challenges of power consumption that accompany the relentless pursuit of Moore's Law. Photonic computing promises a paradigm shift, offering a path beyond the limitations of classical CMOS approaches and opening doors to unprecedented speed and efficiency in information processing, making it a compelling candidate for the next era of advanced computing architectures [24, 25].

Silicon photonics has become a mainstream technology driven by advances in optical communications, leading to millions of integrated photonic devices being produced, mainly as transceivers for data centers. However, to increase the proliferation to billions of units shipped for exciting new applications like sensing and computing, certain challenges need to be addressed [26]. Silicon–Germanium (SiGe) finds widespread use in photonics due to its compatibility with existing silicon technology, enabling seamless integration with traditional electronics on a single chip. This compatibility streamlines manufacturing processes and reduces overall system complexity. Moreover, SiGe allows for precise bandgap engineering, facilitating the design of materials with tailored optical properties suitable for various photonics applications. Its high carrier mobility ensures fast operation of photonic devices, essential for high-speed data transmission in optical communication systems. Additionally, leveraging the established silicon manufacturing infrastructure makes SiGe photonics cost-effective, conducive to large-scale production and widespread adoption. Furthermore, SiGe's compatibility with complementary metal–oxide–semiconductor (CMOS) processes enables the monolithic integration of photonics and electronics, further enhancing efficiency. Its good thermal conductivity properties aid in effective thermal management, ensuring device reliability and performance.

Fig. 1.6 Multi-core technology

1.4   Work Around Moore's Law: Architectural-Level Trends

1.4.1   Multi-cores Technology

Multicore processing can increase performance by running multiple applications concurrently. The decreased distance between cores on an integrated chip enables shorter resource access latency and higher cache speeds when compared to using separate processors or computers [27]. In 1974, Dennard observed that voltage and current should be proportional to the linear dimensions of a transistor. Power is proportional to transistor area and, as transistors get smaller, power density increases. Since around 2006, Dennard's Law has limited processor clock rate to 4 GHz or less. With the end of Moore's Law, Dennard's Law has already limited processor clock rate. i.e., power dissipation limits the clock rates. So, increasing the number of cores can mitigate of the limitation of increasing clock rates due to power dissipation (Fig. 1.6).

1.4.2   Distributed Memory and Computing

The advantage of distributed shared memory is that it offers a unified address space in which all data can be found. The advantage of distributed memory is that it excludes race conditions, and that it forces the programmer to think about data distribution. Distributed computing can help improve performance by having each computer in a cluster handle different parts of a task simultaneously. Scalability. Distributed computing clusters are scalable by adding new hardware when needed [28]. Distributed memory and computing is a technique used in parallel computing to solve large and complex problems by dividing the task into smaller sub-tasks that can be executed in parallel across multiple processors or computers. In distributed memory and computing, each processor or computer has its own local memory, which is separate from the memory of the other processors or computers. To share data between processors or computers, the data must be explicitly transferred from one memory to another using communication protocols such as message passing.

Distributed memory and computing are commonly used in high-performance computing applications, such as scientific simulations, where large amounts of data need to be processed in parallel. By distributing the computation across multiple processors or computers, distributed memory and computing can greatly improve the speed and efficiency of the computation. However, distributed memory and computing also requires careful design and programming to ensure that the different processors or computers work together effectively and that data is transferred efficiently and accurately. Specialized programming languages and tools, such as MPI (Message Passing Interface), are often used to help manage the complexity of distributed memory and computing.

1.4.3   In-Memory Computing

In a conventional von Neumann architecture, data must be transferred between the processor and memory, incurring latency and energy consumption. In-Memory Computing aims to overcome these limitations by performing computations directly within the memory, reducing the need for data movement. This architecture has the potential to improve performance and energy efficiency, particularly for memory-intensive tasks.

1.4.4   HW/SW Co-design

Hardware/Software (HW/SW) co-design offers a powerful approach to optimizing power consumption in computing systems. By integrating hardware and software design efforts, developers can employ various strategies to enhance energy efficiency. One key avenue involves crafting power-aware algorithms that execute tasks with minimal energy expenditure. These algorithms prioritize low-power data processing methods, reduce data movement, and optimize computationally intensive operations. Dynamic power management techniques further enhance efficiency by dynamically adjusting hardware parameters and software execution based on workload and power demands. This includes dynamically scaling CPU frequency and voltage, selectively activating hardware components, and adjusting software execution paths to match system load. Hardware acceleration presents another opportunity, enabling the offloading of intensive tasks from the CPU to specialized hardware accelerators. Low-power hardware design principles are also crucial, encompassing transistor-level optimizations, power gating techniques, and utilization of low-power components. Additionally, power-aware compilation techniques optimize software code for reduced power consumption, while system-level optimizations minimize energy usage across the entire platform. These optimizations encompass communication protocols, data movement, and system-level power management policies. Finally, energy-aware task scheduling dynamically allocates tasks across hardware resources based on energy profiles, deadlines, and resource availability. By employing these strategies in HW/SW co-design, developers can achieve significant improvements in power efficiency across a wide range of computing platforms, from embedded systems to mobile devices and IoT devices, ultimately extending battery life and reducing environmental impact.

1.5   Work Around Moore's Law: Circuit-Level/Logic-Level Trends

1.5.1   Adiabatic Logic

The word ADIABATIC comes from a Greek word that is used to describe thermodynamic processes that exchange no energy with the environment and therefore, no energy loss in the form of dissipated heat. In real-life computing, such an ideal process cannot be achieved because of the presence of dissipative elements like resistances in a circuit. However, one can achieve very low energy dissipation by slowing down the speed of operation and only switching transistors under certain conditions. The signal energies stored in the circuit capacitances are recycled instead of being dissipated as heat. The adiabatic logic is also known as ENERGY RECOVERY CMOS. To reduce the energy dissipation in conventional CMOS, either the supply voltage VDD has to be reduced, or the load capacitance C.

In contrast to the conventional CMOS circuit, the supply voltage in adiabatic case is not constant, but time variable with a slow rising time $(T \gg RC)$. The dissipated energy is smaller than for the conventional case, if the charging time T is larger than 2RC. That is, the dissipated energy can be made arbitrarily small by increasing the charging time. Also, the dissipated energy is proportional to R, as opposed to the conventional case, where the dissipation depends on the capacitance and the voltage swing. Thus, reducing the on-resistance of the PMOS network will reduce the energy dissipation. Adiabatic switching circuits require non-constant, non-standard power supply with time varying voltage. This supply is called pulse power supplies.

To convert a conventional CMOS logic gate into an adiabatic gate, the pull-up and the pull-down networks must be replaced with complementary transmission-gate (T-gate) networks. The T-gate network implementing the pull-up function is used to drive the true output of the adiabatic gate, while the T-gate network implementing the pull-down function drives the complementary output node. Note that all the inputs should also be available in complementary form. Both the networks in the adiabatic logic circuit are used to charge-up as well as charge-down the output capacitance, which ensures that the energy stored at the output node can be retrieved by the power supply, at the end of each cycle. To allow adiabatic operation, the DC voltage source of the original circuit must be replaced by a pulsed-power supply with the ramped voltage output. To convert an Adiabatic CMOS logic gate into a conventional gate, replace each of the PMOS and NMOS devices in the pull-up and pull-down networks with T- gates. Use the pull-up network to drive the true output and the pull-down network to drive the complementary output. Both networks are used to charge and discharge the load capacitance [29–31].

1.5.2 MTCMOS

Multi-threshold CMOS (MTCMOS) circuits reduce standby leakage power with low delay overhead. Most MTCMOS designs cut off the power to large blocks of logic using large sleep transistors [32]. Multi-threshold CMOS (MTCMOS) is a technique used in digital circuit design to reduce power consumption by selectively applying different power supply voltages to different parts of the circuit. In traditional CMOS circuits, all transistors operate at the same voltage level, even if they do not need to operate at the same speed or power level. MTCMOS circuits, on the other hand, use different threshold voltages for different transistors, allowing them to operate at different speeds and power levels. MTCMOS circuits typically have multiple power domains, each of which has a different power supply voltage level. During operation, power is applied only to the domains that are needed, and the other domains are shut off to conserve power. For example, if a particular section of the circuit is not currently being used, its power supply voltage can be reduced or turned off entirely, reducing its power consumption. MTCMOS circuits can also incorporate power gating, which involves adding additional transistors that allow parts of the circuit to be powered on or off as needed. When a section of the circuit is not in use, its power supply can be shut off completely, reducing both static and dynamic power consumption.

1.5.3 Multiple-V_{dd}

With a multi-voltage supply (multi-VDD) approach, some blocks use lower supply voltages than others, creating voltage "islands." This flow gets even more complex when dynamic voltage scaling is used to change the supply voltage level during operation [33]. Reducing power naturally enhances power performance. In traditional digital circuits, all logic gates and transistors operate at the same voltage level, regardless of whether they need to operate at full power or can operate at a lower voltage. Multiple-V_{dd} circuits, on the other hand, divide the circuit into different regions and apply different voltages to each region depending on its power and performance requirements. The basic idea of Multiple-V_{dd} is to use the lowest possible voltage required for a particular logic gate or transistor to function properly. In other words, circuits are designed to operate at the minimum required voltage to maintain functionality. Multiple-V_{dd} circuits typically have two or more power domains, each with a different power supply voltage level. During operation, power is applied only to the domains that are needed, and the other domains are shut off to conserve power. By applying the lowest possible voltage to each domain, Multiple-V_{dd} circuits can reduce both dynamic and static power consumption. Multiple-V_{dd} is commonly used in modern digital circuits, particularly in low-power applications such as mobile devices, where power consumption is a critical factor. By using multiple power supply voltages throughout the circuit, Multiple-Vdd circuits can reduce power consumption while still meeting performance requirements.

1.5.4   Clock-Gating

Clock Gating (CG) which switches off certain Flip-Flops (FFs) when it is found that the input to a Flip-Flop from a previous combinational cloud has not changed. This allows for dynamic power savings down the logic chain as subsequent FFs also can be gated. Clock Gating can only reduce dynamic power consumption as leakage power consumption remains unchanged whether a certain circuit block is gated or not gated as the individual transistors in the block are still connected to the power grid [34]. In traditional digital circuits, the clock signal is applied to all parts of the circuit, even if they are not currently in use. This means that the circuit is constantly switching and consuming power, even when it is not actually performing any useful work. Clock gating involves adding additional logic gates to the circuit that allow the clock signal to be selectively applied to individual sections of the circuit, depending on whether they are currently needed or not. When a section of the circuit is not in use, its clock signal can be disabled, which stops the circuit from switching and conserves power. When the section is needed again, the clock signal is enabled and the circuit resumes operation. Clock gating is particularly effective in reducing dynamic power consumption, which is the power consumed by the circuit when it is actively switching. It is commonly used in modern digital circuits, particularly in low-power applications such as mobile devices, where power consumption is a critical factor.

1.5.5   Power-Gating

The need to reduce standby leakage power led to the introduction of Power Gating. Power gating is a technique used in integrated circuit design to reduce power consumption, by shutting off the current to blocks of the circuit that are not in use [34]. In traditional digital circuits, all transistors and logic gates are powered on at all times, even if they are not currently in use. Power gating involves adding additional transistors to the circuit that allow individual sections of the circuit to be powered on or off as needed. When a section of the circuit is not in use, its power supply can be shut off completely, reducing both dynamic and static power consumption. When the section is needed again, the power supply can be turned back on and the circuit resumes operation. Power gating can be used in conjunction with other low-power design techniques, such as voltage scaling and clock gating, to further reduce power consumption in digital circuits. It is particularly effective in reducing static power consumption, which is power consumed by the circuit even when it is not actively switching. Power gating is commonly used in modern digital circuits, particularly in low-power applications such as mobile devices, where power consumption is a critical factor. By selectively shutting off power to unused parts of the circuit, power gating can significantly reduce power consumption while still meeting performance requirements.

1.5.6   Asynchronous Circuits

In an asynchronous circuit the next computation step can start immediately after the previous step has completed: there is no need to wait for a transition of the clock signal. This leads, potentially, to a fundamental performance advantage for asynchronous circuits, an advantage that increases with the variability in delays associated with these computation steps [35]. Asynchronous circuits are a type of digital circuit design where the individual components (gates, flip-flops, etc.) operate independently of a clock signal. In contrast, synchronous circuits operate on a clock signal that is used to synchronize the timing of all the components in the circuit. In asynchronous circuits, each component operates based on its inputs and outputs, without regard for the timing of other components in the circuit. This makes asynchronous circuits more flexible and adaptable to changes in the input signals, as they can respond to changes as they occur rather than waiting for the next clock cycle. One of the main advantages of asynchronous circuits is their low power consumption. Synchronous circuits require a clock signal that must be distributed throughout the circuit, resulting in higher power consumption due to the clock distribution network. Asynchronous circuits, on the other hand, do not require a clock signal and can operate on demand, leading to lower power consumption. However, asynchronous circuits can be more difficult to design and analyze compared to synchronous circuits, as the behavior of each component depends on the timing of its inputs and outputs, which can be difficult to predict. As a result, designing and testing asynchronous circuits can be more time-consuming and complex. Asynchronous circuits are commonly used in applications where low power consumption and high flexibility are important, such as in mobile devices and sensor networks.

1.5.7   Pipelining

Pipelining allows processing tasks in parallel. Creating parallel operators to process events improves efficiency (Fig. 1.7) [36]. Pipelining is a technique used in computer architecture to increase the performance of processors by breaking down the execution of instructions into a series of smaller, independent stages, allowing for the concurrent processing of multiple instructions. In pipelining, the processor is divided into a series of stages, with each stage performing a specific task in the instruction execution process. When an instruction is fetched from memory, it is then passed to the first stage of the pipeline, which might be an instruction decoder. Once the instruction has been decoded, it is passed to the next stage, which might be an arithmetic unit or a memory access unit, and so on. While an instruction is being processed in a particular stage, the next instruction can be fetched and start processing in the first stage. This overlap of processing allows the processor to perform multiple instructions at the same time, thereby increasing its throughput and improving its overall performance. Pipelining can be further optimized through techniques such as instruction-level parallelism, which involves breaking down instructions into smaller, independent operations that can be executed in parallel, and branch prediction, which involves predicting the outcome of a conditional branch instruction before it is executed to minimize pipeline stalls caused by incorrect predictions. Overall, pipelining is a powerful technique for improving processor performance by dividing the instruction execution process into smaller, independent stages that can be executed concurrently, allowing for faster processing of instructions and increased throughput.

1.5.8   Data-Encoding

The data encoding approach is the most promising technique to decrease the dynamic power dissipation on on-chip data interconnects and as a result overall device performance improve. The Data encoding techniques aim to reduce the switching activity factor which directly affects the dynamic power dissipation and cross talk delay during transmission on interconnects. Low power data encoding techniques intend to convert the data such that the self-switching activity and coupling switching activity on interconnects are decreased. Gray code was the first data encoding technique which utilized the principle of only one transition per every transmitted address. Hence, Gray code provides less switching activity without any redundancy [37].

Time (in clock cycles)

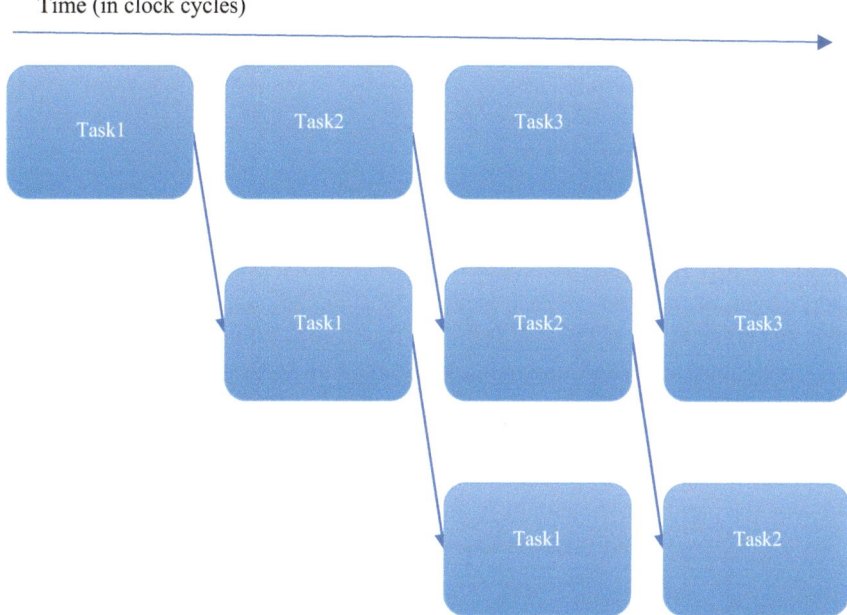

Fig. 1.7 An illustrative diagram for pipelining

1.5.9 Repeater

A repeater is an electronic device in a communication channel that increases the power of a signal and retransmits it, allowing it to travel further. Hence, it improves the performance. Repeater insertion in interconnects is an increasingly crucial element in the physical design of high-performance VLSI systems. Interconnect tuning and repeater insertion are necessary to optimize interconnect delay, signal performance and integrity, and interconnect manufacturability and reliability [38].

1.6 Work Around Moore's Law: Software-Level/OS-Level Trends

1.6.1 Concurrency

In a 1967, Amdhal argued that the fraction of a computation that is not parallelizable is significant enough to favor single-processor systems. He reasoned that large-scale computing capabilities can be achieved by enhancing the performance of single processors, rather than building multiprocessor systems. Having concurrency allows the operating system to run multiple applications at the same time. This results in better resource utilization, better average response time, and better performance [39].

1.6.2 Partitioning

A partition is a logical division of a hard disk that is treated as a separate unit by the operating system. In operating systems, Memory Management is the function responsible for allocating and managing a computer's main memory. Memory Management function keeps track of the status of each memory location, either allocated or free to ensure effective and efficient use of Primary Memory. There are two Memory Management Techniques: Contiguous, and Non-Contiguous. In Contiguous Technique, the executing process must be loaded entirely in the main memory. Contiguous Technique can be divided into Fixed (or static) partitioning, Variable (or dynamic) partitioning. Dynamic partitioning helps in overcoming the difficulties caused by the process of fixed partitioning. In dynamic partitioning, the partition size initially is not declared. It is declared at the moment of process loading. The very first partition has to be reserved for the OS. The left space gets divided into various parts. The actual size of every partition would be equal to the process size. The size of the partition varies according to the requirement of the process. This way, internal fragmentation can be easily avoided [40].

1.6.3 Sleep Mode

Sleep mode, sometimes called standby or suspend mode, is a power-sparing state that a computer can enter when not in use. The computer's state is maintained in random access memory (RAM). In sleep mode, since the data is stored in RAM, the resumption is immediate and no time is lost. But a hibernating system needs comparatively more time to resume as it needs time to read back the data from the hard disk or other permanent memory storage. Power consumption: Lower in hibernate mode.

1.7 Work Around Moore's Law: Packaging-Level Trends

Microfluidic Cooling is used to implement microfluidic channels within the packaging to circulate coolant and efficiently remove heat from hotspots. Microfluidic Cooling offers a promising solution to the thermal challenges posed by increasingly dense and power-hungry electronic systems. This technology not only enhances thermal management capabilities but also contributes to the overall efficiency and longevity of electronic devices.

References

1. Salah, K., El Rouby, A., Ragai, H., & Ismail, Y. (2011). TSVs macro-modeling framework. In *2011 International Conference on Energy Aware Computing* (pp. 1–4). IEEE.
2. Roy, K., Jung, B., & Raghunathan, A. (2010). Integrated systems in the more-than-Moore era: designing low-cost energy-efficient systems using heterogeneous components. In *23rd International Conference on VLSI Design.*
3. Lu, Z., & Jantsch, A. (2009). Trends of terascale computing chips in the next ten years. In *ASICON.*
4. Ionescu, A. M. (2009). Nanoelectronics roadmap: Evading Moore's law" EWME.
5. Rairigh, D. (2005). *Limits of CMOS technology scaling and technologies beyond-CMOS.* IEEE.
6. Wang, M. C. Independent-gate FinFET circuit design methodology. *IAENG International Journal of Computer Science.*
7. Zhang, Y. (2020). The application of third generation semiconductor in power industry. *E3S Web of Conferences, 198*, 04011. https://doi.org/10.1051/e3sconf/202019804011
8. Weerasekera, R., Grange, M., Pamunuwa, D., Tenhunen, H., & Zheng, L.-R. (2009). Compact modelling of through-silicon vias (TSVs) in three dimensional (3-D) integrated circuit. In *Proceedings of the IEEE International Conference on 3D System Integration (3D IC)*, San Francisco, USA.
9. Iyer, S. S. (2018). Three-dimensional integration: An industry perspective. *Materials Research Society Bulletin, 40*(3), 225–232.
10. Salah, K., et al. (2011). TSVs macro-modeling framework. In *2011 International Conference on Energy Aware Computing.* IEEE.
11. Carloni, L. P., Xie, Y. (2009). *Networks-on-chip in emerging interconnect paradigms: Advantages and challenges.* IEEE.
12. Monemi, A., Pérez, I., Leyva, N., Vallejo, E., Beivide, R., & Moretó, M. (2021). PlugSMART: A pluggable open-source module to implement multihop bypass in networks-on-chip. In *2021 15th IEEE/ACM International Symposium on Networks-on-Chip (NOCS)* (pp. 41–48).
13. Bergman, K., Carloni, L. P., Kash, J. A., & Vlasov, Y. (2009). *On-chip photonic communication for high-performance multi-core processors.* IEEE.
14. Lau, J. H. (2022). Recent advances and trends in advanced packaging. *IEEE Transactions on Components, Packaging, and Manufacturing Technology*, 228–252.
15. Salah, K., et al. (2012). Effect of non-uniform substrate doping profile on the electrical performance of through-silicon-via for low power application. In *2012 International Conference on Energy Aware Computing.* IEEE.
16. Japa, A., Sahoo, S., Vaddi, R., & Kaushik, B. K. (2022). Hardware security exploiting post-CMOS devices: Fundamental device characteristics, state-of-the-art countermeasures, challenges and roadmap. *IEEE Circuits and Systems Magazine, 21*, 4–30. https://doi.org/10.1109/MCAS.2021.3092532
17. http://www.physorg.com/news90607516.html
18. Hassan, S., Asghar, M. (2010). Limitation of silicon based computation and future prospects. In *Second International Conference on Communication Software and Networks.*
19. Kaiser, J., & Datta, S. (2021). Probabilistic computing with p-bits. *Applied Physics Letters, 119*(15).
20. Mohamed, K. S. (2020). Quantum computing and DNA computing: Beyond conventional approaches. In *Neuromorphic computing and beyond.* Springer. https://doi.org/10.1007/978-3-030-37224-8_7
21. Xie, Y. (2014). *Emerging memory technologies.* Springer.

22. Yu, S. (2016). *Resistive random access memory (RRAM)*. Morgan & Claypool Publishers.
23. El Srouji, L., Krishnan, A., Ravichandran, R., Lee, Y., On, M., Xiao, X., & Ben Yoo, S. J. (2022). Photonic and optoelectronic neuromorphic computing. *APL Photonics 7*(5).
24. Margalit, N., Xiang, C., Bowers, S. M., Bjorlin, A., Blum, R., & Bowers, J. E. (2021). Perspective on the future of silicon photonics and electronics. *Applied Physics Letters, 118*(22), 220501.
25. Shastri, B. J., Tait, A. N., Lima, T., Pernice, W. H. P., Bhaskaran, H., Wright, C. D., & Prucnal, P. R. (2021). Photonics for artificial intelligence and neuromorphic computing. *Nature Photonics, 15*(2), 102–114.
26. Shekhar, S., Bogaerts, W., Chrostowski, L., Bowers, J. E., Hochberg, M., Soref, R., & Shastri, B. J. (2024). Roadmapping the next generation of silicon photonics. *Nature Communications, 15*(1), 751.
27. https://insights.sei.cmu.edu/blog/multicore-processing/
28. https://www.techtarget.com/whatis/definition/distributed-computing
29. Mandavilli, S., & Paramahans, P. (2009). An efficient adiabatic circuit design approach for low power applications. *International Journal of Recent Trends in Engineering, 2*(1).
30. Anuar, N., Takahashi, Y., & Sekine, T. (2010). *Adiabatic logic versus CMOS for low power applications*. IEEE.
31. Salah, K., & Ismail, Y. (2015). Design of adiabatic TSV, SWCNT TSV, and air-gap coaxial TSV. In *2015 IEEE International Symposium on Circuits and Systems (ISCAS)*, Lisbon, Portugal, 2015 (pp. 1953–1956). https://doi.org/10.1109/ISCAS.2015.7169056
32. Calhoun, H., Honore, F. A., & Chandrakasan, A. P. (2004). A leakage reduction methodology for distributed MTCMOS. *IEEE Journal of Solid-State Circuits, 39*(5), 818–826. https://doi.org/10.1109/JSSC.2004.826335
33. https://www.electronicdesign.com/power-management/article/21760077/lowpower-design-with-multivdd-flows
34. Patil, V. C. (2004). *Effect of clock and power gating on power distribution network noise in 2D and 3D integrated circuits noise in 2D and 3D integrated circuits*. Master Thesis.
35. http://www1.cs.columbia.edu/~nowick/async-applications-PIEEE-99-berkel-josephs-nowick-published.pdf
36. https://www.techtarget.com/whatis/definition/pipelining
37. Prasad, T., Chennakesavulu, M., Prasad, T., & Sumalatha, V. (2018). Data encoding techniques to improve the performance of system on chip. *Journal of King Saud University—Computer and Information Sciences, 34*. https://doi.org/10.1016/j.jksuci.2018.12.003
38. Muddu, S., Sarto, E., Hofmann, M., & Bashteen, A. (1998). Repeater and interconnect strategies for high-performance physical designs. In *Proceedings. XI Brazilian Symposium on Integrated Circuit Design* (Cat. No. 98EX216), Rio de Janeiro, Brazil, 1998 (pp. 226–231). https://doi.org/10.1109/SBCCI.1998.715447
39. https://eng.libretexts.org/Courses/Delta_College/Operating_System%3A_The_Basics/05%3A_Process_Synchronization/5.1%3A_Introduction_to_Concurrency
40. https://byjus.com/gate/dynamic-partitioning-in-operating-system-notes/

EDA Tools Modeling Methods

2

"Efficient Computational Methods"

2.1 Introduction

Many physical phenomena can be described mathematically by the same class of system. Any system can be represented by a set of continuous Partial Differential Equations (PDEs) or discrete Ordinary Differential Equations (ODEs). At the same time, any set of PDEs should be transformed into a system of ODEs which can be linear ODEs or nonlinear ODEs. So, discretization is needed which approximates the behavior of the continuous systems. For example, Maxwell's equations in electromagnetic describe the behavior of the system continuously in time and space [1]. Most CAD tools use the numerical Finite Element Method (FEM) approximation to accurately discretize in space, model and simulate these continuous structure-level VLSI systems. FEM is considered a standard approach effectively applied for solving structural, fluid, and multi-physics problems numerically.

Structural simulation deals with the mechanical behavior of structures and their response to external forces. Structural simulation refers to the process of using computer-based models to simulate the behavior and response of physical structures under various conditions. It is widely used in engineering and design fields to analyze the structural integrity, performance, and safety of different types of structures, such as buildings, bridges, vehicles, and machines. Structural simulation involves creating a virtual model of the structure using specialized software. This model typically includes the geometry, material properties, and boundary conditions of the structure. The software then uses mathematical equations and numerical methods to predict how the structure will behave when subjected to various loads, forces, or environmental conditions. The simulation can provide insights into factors such as stress distribution, deformation, vibration, fatigue life, and failure modes. It can help engineers evaluate different design alternatives, assess the impact of changes or modifications, optimize structural performance, and

© The Author(s), under exclusive license to Springer Nature Switzerland AG 2025
K. S. Mohamed, *Next Generation EDA Flow*, Synthesis Lectures on Engineering, Science, and Technology, https://doi.org/10.1007/978-3-031-88435-1_2

ensure compliance with safety standards and regulations. There are different types of structural simulation techniques, including finite element analysis (FEA), computational fluid dynamics (CFD), and multibody dynamics (MBD). These techniques allow engineers to analyze different aspects of the structure, such as static and dynamic behavior, thermal effects, fluid–structure interaction, and more. Structural simulation has become an essential tool in the design and analysis process, enabling engineers to make informed decisions, improve efficiency, reduce costs, and enhance the overall performance and reliability of structures.

Solving linear ODEs results in matrix form system that can be solved using direct method such as Gaussian elimination method or **indirect** method (iterative methods) such as Jacobi-method and solving nonlinear ODEs can be done by newton's method. These methods are useful for moderate size problems.

While direct solvers offer accuracy and reliability, they tend to demand significant computational resources and memory, making them less efficient for large-scale problems due to their cubic scaling with matrix size. In contrast, iterative solvers refine an initial guess iteratively to approximate the solution. Iterative solvers are often more memory-friendly and computationally efficient for large systems, as they only need to store and manipulate sparse matrices and vectors. Scalars, vectors, and matrices are fundamental mathematical entities with distinct characteristics and applications. A scalar is a single quantity with magnitude but no direction, such as time, temperature, or mass. Vectors possess both magnitude and direction. Matrices are arrays of numbers arranged in rows and columns (Table 2.1).

Solving high-order (system complexity is referred to as the order of the system) or high-degree of freedom of these discretized differential equations is a time-consuming process. Model order reduction (MOR) technique is a compression method to reduce the order of the full-order ODEs for fast computations and less storage requirements and at the same time it keeps the same characteristics of the full system and it should be passivity-guaranteed and stability-guaranteed model [2]. There should be a global error bound between the transfer functions of the original and reduced/compact systems. MOR is developed after the FEM discretization or any other discretization method to reduce the matrix size as illustrated in Fig. 2.1 [3–10].

Any physical system can be represented mathematically by linear or nonlinear partial differential equations (PDEs). PDE involves two or more independent variables and it is used to represent 2D or 3D problems [11]. When a function involves one independent variable, the equation is called an ordinary differential equation (ODE) and it is used to represent 1D problems. The discretization of partial differential equations leads to a large sparse or dense system of linear equations (SLEs) or system of non-linear equations (SNLEs) as depicted in Fig. 2.1. Sometimes the number of theses equations can reach 100 million or more [11].

Table 2.1 Comparison between scalar, vector and matrix

	Scalar	Vector	Matrix
Definition	Single quantity with magnitude and no direction	Quantity with magnitude and direction	Array of numbers arranged in rows and columns
Dimension	0-dimensional (no length, area, volume, etc.)	1-dimensional (length)	2-dimensional (rows and columns)
Notation	Typically represented by lowercase letters or Greek symbols	Typically represented by bold lowercase letters or lowercase letters with arrows on top	Typically represented by uppercase letters or bold uppercase letters
Examples	$3, -2.5, \pi$	$[2, -1, 0], (4, -3), \langle 1, 2, -3\rangle$	$[[1, 2, 3], [4, 5, 6], [7, 8, 9]]$
Operations	Addition, subtraction, multiplication, division, exponentiation	Addition, subtraction, scalar multiplication, **dot product, cross product**	Addition, subtraction, scalar multiplication, matrix multiplication, **matrix inverse**
Size/Shape	N/A (scalar has no size or shape)	Length or magnitude determines the size	Number of rows and columns determine the size
Representation	Represented by a single value	Represented by a column or row of values	Represented by a grid of values
Transformation	No transformation operations	Translation, rotation, scaling, projection	Linear transformations, eigenvalue decomposition, diagonalization
Application	Scalar quantities such as time, temperature, mass	Displacement, velocity, force in physics; RGB color in computer graphics	Linear systems, transformations, data analysis, computer graphics

In general, methods to solve PDEs and ODEs can be classified into two categories: analytical methods and iterative methods. Due to the increasing complexities encountered in the development of modern VLSI technology, analytical solutions usually are not feasible. So, iterative methods are used.

SNLEs are generally represented in $f(x) = 0$ form. A SNLEs solver must determine the values of x. Non-linear equations can be in the form of cos, sin, log, polynomial function, parabolic function, exponential, and complex. They cover different domains such as materials, fluid, chemistry and physics. In general, methods to solve SNLEs are iterative methods [11].

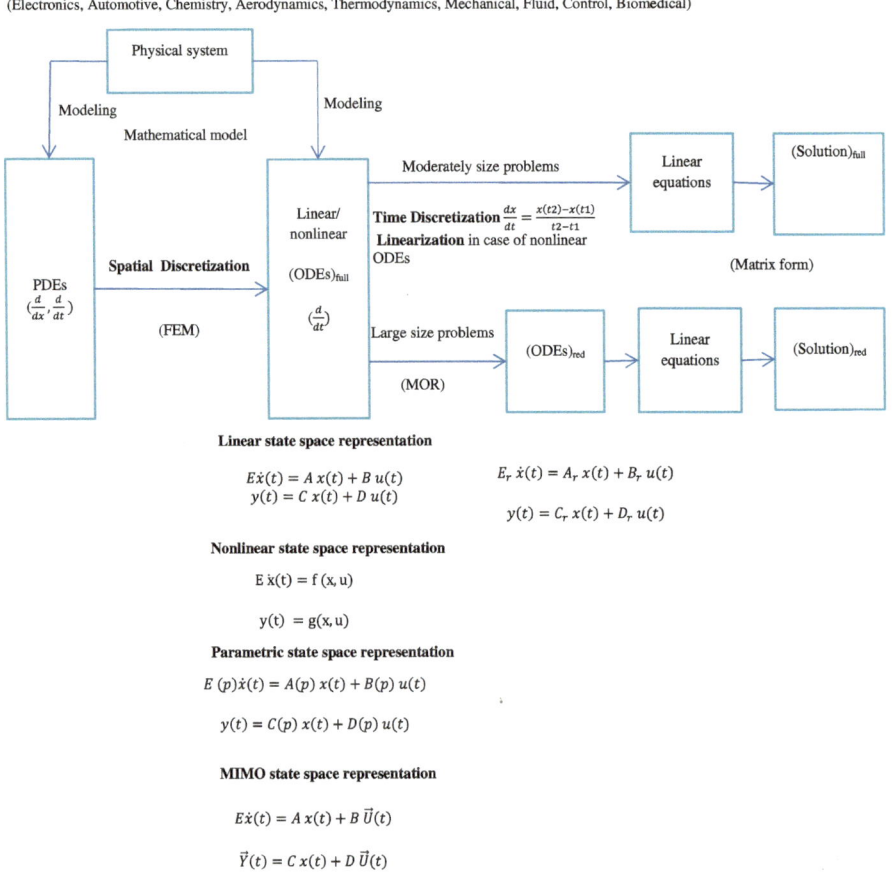

(Electronics, Automotive, Chemistry, Aerodynamics, Thermodynamics, Mechanical, Fluid, Control, Biomedical)

Linear state space representation

$$E\dot{x}(t) = A\,x(t) + B\,u(t)$$
$$y(t) = C\,x(t) + D\,u(t)$$

$$E_r\,\dot{x}(t) = A_r\,x(t) + B_r\,u(t)$$
$$y(t) = C_r\,x(t) + D_r\,u(t)$$

Nonlinear state space representation

$$E\,\dot{x}(t) = f\,(x, u)$$

$$y(t) = g(x, u)$$

Parametric state space representation

$$E\,(p)\dot{x}(t) = A(p)\,x(t) + B(p)\,u(t)$$

$$y(t) = C(p)\,x(t) + D(p)\,u(t)$$

MIMO state space representation

$$E\dot{x}(t) = A\,x(t) + B\,\vec{U}(t)$$

$$\vec{Y}(t) = C\,x(t) + D\,\vec{U}(t)$$

Fig. 2.1 Typical mathematical modeling flow for complex physical systems: the big picture. MOR is needed for large size problems, given an ODE of order n, find another ODE of order r, where r ≪ n with "essentially" the same "properties" and stability, passivity are preserved. y(t) is the output, u(t) is the input, x(t) the state. We replace the set of ODEs by a smaller set of ODEs without sacrificing the accuracy of the system behavior. The universe is continuous, but computers are discrete. A computer does not understand Physics/Math equations, so we need to discretize/sample them to be able to program them [17]

The most common iterative methods are Newton, Quai-Newton, Secant and Muller. These methods are summarized in Fig. 2.1 [12].

SLEs are generally represented in a matrix form. Given a system of linear equations in the form of $Ax = b$, where x is the unknowns, there can be one solution, an infinite number of solutions or no solution. A SLEs solver must determine the values of x for which the product with matrix (A) generates the vector constant (b). In general, methods to solve SLEs can be classified into two categories: direct methods and iterative methods. Both iterative and direct solvers are widely used [13–15].

Direct methods determine the exact solution in a single process while iterative methods start with an initial guess and compute a sequence of intermediate results until it converges to a final solution. However, convergence is not always guaranteed for iterative methods since they are more sensitive to the matrix type being solved (Sparse, Dense, Symmetric, Square) which limits the scope of the range of the matrices that can be solved.

But, iterative methods are useful for finding solutions to large systems of linear equations where direct methods are prohibitively expensive in terms of computation time.

Direct methods are typically used for dense matrices of small to moderate size, which consist mainly of nonzero coefficients, as they would require large number of iterations and thus more memory accesses, if iterative methods are used. On the other hand, iterative methods are preferred for sparse matrices or dense matrices of large size, which are matrices that have a lot of zero coefficients, to make use of this special structure in reducing the number of iterations.

The most common direct methods are Gaussian elimination, Gauss-Jordan elimination and LU de-composition. All of them have a computational complexity of order n^3 and are sensitive to numerical errors. The most common iterative methods are Jacobi, Gauss Seidel, and Conjugate Gradient methods. They have computational complexity of order n^2 for each iteration step. These methods are summarized in Fig. 2.1.

Recently, advanced methods based on hybrid solutions and machine learning is proposed to solve SNLEs/SLEs. Hybrid methods combine new techniques with the conventional iterative methods.

ML-based methods such as genetic algorithm (GA) and artificial neural network (ANN) enhance solving process by accelerating the convergence of iterative numerical methods which reduces the computational efforts [16] (Fig. 2.2).

2.2   Numerical Analysis for Electronics

Numerical analysis, a branch of mathematics, employs numerical approximation to address continuous problems, offering approximate yet accurate numerical solutions. This is especially useful when obtaining exact solutions is impractical or too costly. Techniques such as finite difference methods, error propagation, and interpolation fall within the scope of numerical analysis. Given the complexity of many scientific and technical applications, numerical approaches become necessary since analytical solutions are often unattainable. The primary objectives of numerical analysis are to provide precise approximations of mathematical problem solutions and to develop efficient algorithms for their rapid resolution. With the advancement of computational technology, numeracy has become indispensable in modern science and engineering. Software programs like Matlab, Mathematica, and Maple have emerged to simplify complex problem-solving

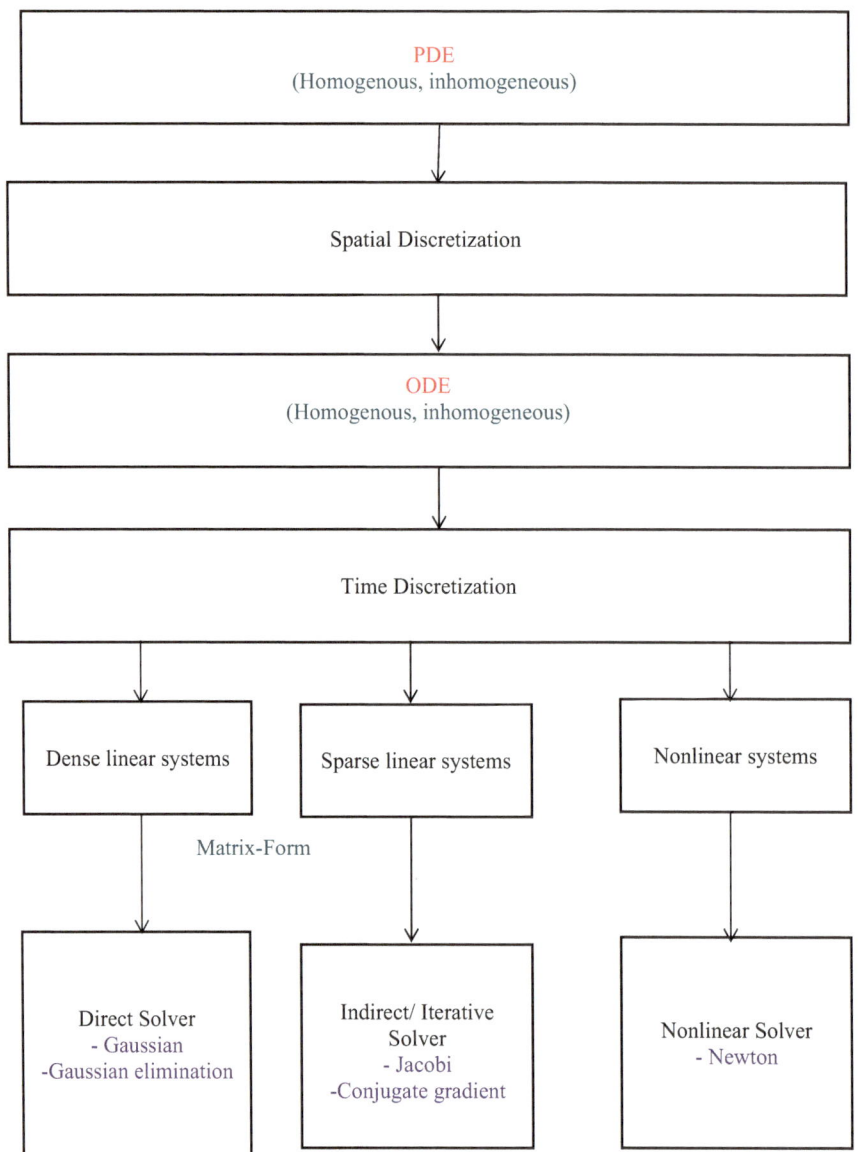

Fig. 2.2 Different methods of solving systems of linear equations: the big picture

tasks. These programs utilize standard numeric methods, enabling users to obtain results swiftly without delving into intricate numerical details. Numerical analysis primarily finds application in mathematics and computer science to address numerical challenges encountered in real-world scenarios, spanning domains such as scientific research, engineering, healthcare, and business.

Since the advent of integrated circuits (ICs), numerical analysis has played an important role in simulating analog, mixed-signal and RF designs. Circuit simulations, electromagnetic (EM) simulations, device simulations are built using many numerical analysis methods [18–25]. The universe is continuous, but computers are discrete. Computers do not understand Physics/Maths equations, so we need to discretize/sample them to be able to program them.

2.2.1   Why EDA?

Electronic Design Automation (EDA) is a set of software tools used by electronic designers to design and analyze electronic systems. EDA is important because it helps designers create complex electronic systems with greater efficiency, speed, and accuracy. Electronic systems are becoming more complex with each passing year, and the use of EDA tools is essential to keep up with this complexity. EDA tools help designers simulate and optimize their designs before they are built, reducing the need for costly and time-consuming physical prototypes. EDA tools also enable designers to explore a wider range of design options and quickly iterate through design cycles, leading to faster time-to-market for new products. Additionally, EDA tools help ensure the quality and reliability of electronic systems by detecting potential design flaws and identifying areas for improvement. Overall, EDA plays a critical role in the development of modern electronic systems, and without it, the design and development process would be much slower, less efficient, and less reliable. As complexity of current day electronic design increases, manual design becomes unrealistic. Moreover, automation ensures fast Time to market and fewer errors. VLSI cycle is shown in Fig. 2.3. Numerical methods are used at circuit design level. Logic design uses Booleans [26–30].

2.2.2   Applications of Numerical Analysis

Numerical analysis is a branch of mathematics that deals with the development and application of computational algorithms and methods for solving mathematical problems that cannot be solved analytically. The field of numerical analysis includes many applications such as (Fig. 2.4):

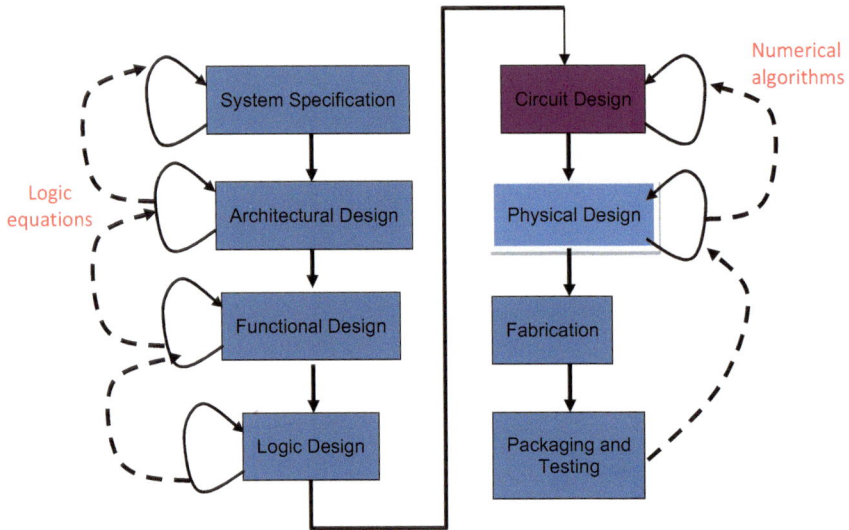

Fig. 2.3 VLSI cycle

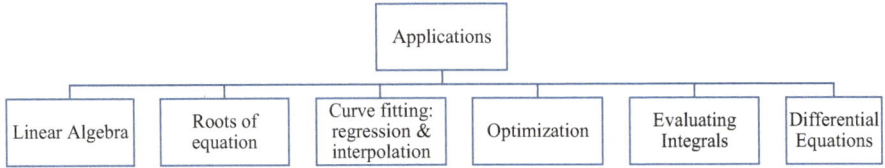

Fig. 2.4 Applications of numerical analysis

- **Computing values of functions**: One of the simplest problems is the evaluation of a function at a given point. The most straightforward approach, of just plugging in the number in the formula is sometimes not very efficient. For polynomials, a better approach is using the Horner scheme, since it reduces the necessary number of multiplications and additions.
- **Interpolation** solves the following problem given the value of some unknown function at a number of points, what value does that function have at some other point between the given points?
- **Extrapolation** is very similar to interpolation, except that now the value of the unknown function at a point which is outside the given points must be found.
- **Regression** is also similar, but it considers that the data is imprecise. Given some points, and a measurement of the value of some function at these points (with an error), the unknown function can be found. The least squares-method is one way to achieve this.

- **Solving equations and systems of equations**: Another fundamental problem is computing the solution of some given equation. Two cases are commonly distinguished, depending on whether the equation is linear or not.
- **Solving eigenvalues or singular value problems**: Several important problems can be phrased in terms of eigenvalue decompositions or singular value decompositions. For instance, the spectral image compression algorithm is based on the singular value decomposition. The corresponding tool in statistics is called principal component analysis.
- **Optimization**: Optimization problems ask for the point at which a given function is maximized (or minimized). Often, the point also has to satisfy some constraints. The field of optimization is further split in several subfields, depending on the form of the objective function and the constraint. For instance, linear programming deals with the case that both the objective function and the constraints are linear. A famous method in linear programming is the simplex method.
- **Evaluating integrals**: Numerical integration, in some instances also known as numerical quadrature, asks for the value of a definite integral. Popular methods use one of the Newton–Cotes formulas (like the midpoint rule or Simpson's rule) or Gaussian quadrature. These methods rely on a "divide and conquer" strategy, whereby an integral on a relatively large set is broken down into integrals on smaller sets. In higher dimensions, where these methods become prohibitively expensive in terms of computational effort, one may use **Monte Carlo** or quasi-Monte Carlo methods, or, in modestly large dimensions, the method of sparse grids. Monte Carlo methods refer to a series of statistical methods essentially used to find solutions to things such as computing the expected values of a function or integrating functions that can't be integrated analytically because they don't have a closed-form solution for example. To run an MC algorithm, we first need to be able to generate random numbers, generally with a given probability distribution). For this reason, the development of algorithms for generating such "random" numbers (these algorithms are called **pseudorandom number generator**), has been an important field of research in computing technology [31].
- **Differential equations**: Numerical analysis is also concerned with computing (in an approximate way) the solution of differential equations, both ordinary differential equations and partial differential equations.

2.2.3 Approximation Theory

Use computable functions $p(x)$ to approximate the values of functions $f(x)$ that are not easily computable or use approximations to simplify dealing with such functions. The most popular types of computable functions $p(x)$ are polynomials, rational functions, and piecewise versions of them, for example spline functions. Trigonometric polynomials are also a very useful choice.

- Best approximations: Here a given function f(x) is approximated within a given finite-dimensional family of computable functions. The quality of the approximation is expressed by a functional, usually the maximum absolute value of the approximation error or an integral involving the error. Least squares approximations and minimax approximations are the most popular choices.
- Interpolation: A computable function p(x) is to be chosen to agree with a given f(x) at a given finite set of points x. The study of determining and analyzing such interpolation functions is still an active area of research, particularly when p(x) is a multivariate polynomial.
- Fourier series: A function f(x) is decomposed into orthogonal components based on a given orthogonal basis $\{\varphi 1, \varphi 2, \dots\}$, and then f(x) is approximated by using only the largest of such components. The convergence of Fourier series is a classical area of mathematics, and it is very important in many fields of application. The development of the Fast Fourier Transform in 1965 spawned a rapid progress in digital technology. In the 1990s wavelets became an important tool in this area.
- Numerical integration and differentiation: Most integrals cannot be evaluated directly in terms of elementary functions, and instead they must be approximated numerically. Most functions can be differentiated analytically, but there is still a need for numerical differentiation, both to approximate the derivative of numerical data and to obtain approximations for discretizing differential equations.

2.3   Different Methods for Solving PDEs and ODEs

Partial differential equations (PDE) and ordinary differential equations (ODE) are used in modeling and solving many problems in various fields such as physics, chemistry, biological, and mechanics (Table 2.2).

Differential equations represent the rate of change in these systems and they are classified according to their **order**. For example, a first order equation includes a first derivative as its highest derivative. Also, it can be **linear** or nonlinear. Some of physical problems are governed by first order PDEs while numerous problems are governed by second order PDEs. Only few problems are governed by higher order PDEs. Higher order equations can be reduced to a system of first order equations, by redefining a variable [32, 33].

Moreover, a linear differential equation is **homogeneous** if every term contains the dependent variable or its derivatives. An **inhomogeneous** differential equation has at least one term that contains no dependent variable (one or more terms involve functions of independent variables or constants).

Table 2.2 Examples for differential equations

Domain	Example	Differential equation
Electromagnetic	Wave equation	Linear $\nabla^2 u \frac{1}{c^2}\frac{\partial^2 u}{\partial t^2}$ Nonlinear $\nabla^2 u \frac{1}{C(c)^2}\frac{\partial^2 u}{\partial t^2}$
	Poisson's equation	$\nabla^2 u \frac{\rho}{\varepsilon_0}$
	Laplace's equation	$\nabla^2 u = 0$
	Maxwell equations	$\nabla \times \vec{E} = -\mu\frac{\partial \vec{H}}{\partial t}$ $\nabla \times \vec{H} = -\varepsilon\frac{\partial \vec{E}}{\partial t} + \sigma \times \vec{E}$ $\nabla \cdot \varepsilon\vec{E} = \rho$ $\nabla \cdot \mu\vec{H} = 0$
Thermodynamics	Diffusion equation	$\nabla^2 u \frac{1}{h^2}\frac{\partial u}{\partial t}$
Quantum mechanics	Schrödinger's equation	$-\frac{\hbar^2}{2m}\nabla^2 u + Vu = i\hbar\frac{\partial u}{\partial t}$
Mechanics	Newton's laws	$F = ma = m\frac{d^2 x}{dt^2}$

The general form for the 2nd order homogeneous partial differential equation is given by:

$$A\frac{\partial^2 u}{\partial x^2} + B\frac{\partial^2 u}{\partial x\partial y} + C\frac{\partial^2 u}{\partial y^2} + D\frac{\partial u}{\partial x} + E\frac{\partial u}{\partial y} + Fu = 0 \qquad (2.1\text{a})$$

The general form for the 2nd order inhomogeneous partial differential equation is given by:

$$A\frac{\partial^2 u}{\partial x^2} + B\frac{\partial^2 u}{\partial x\partial y} + C\frac{\partial^2 u}{\partial y^2} + D\frac{\partial u}{\partial x} + E\frac{\partial u}{\partial y} + Fu = G \qquad (2.1\text{b})$$

where A, B, C, D, E, F, and G are either real constants or real-valued functions of x and/or y. The different types of PDEs and the different solutions for them are summarized in Table 2.3. Both PDE and ODE can be homogenous or inhomogeneous.

There are many methods to solve PDEs and ODEs. We need to specify the solution at $t = 0$, or $x = 0$, i.e., we need to specify the **initial conditions** and the **boundary values**. Analytical methods are based on proposing a trial solution or using separation of variable then integration or choosing a basis set of functions with adjustable parameters and proceed approximating the solution by varying these parameters.

Table 2.3 The different types of PDEs and the different solutions

	PDE type	Solution type	Example
$B^2 - 4AC < 0$	Elliptic	Equilibrium or steady state	Poisson equation
$B^2 - 4AC = 0$	Parabolic	Solution "propagates" or diffuses	Heat equation
$B^2 - 4AC > 0$	Hyperbolic	Solution propagates as a wave	Wave equation

But, due to the increasing complexities encountered in the development of modern VLSI technology, analytical solutions usually are not feasible. So, iterative methods are used. The explanation of the methods will be applied on boundary value problems partial differential equation with Dirichlet conditions.

2.3.1 Iterative Methods for Solving PDEs and ODEs

1. Finite Difference Method (Discretization)

FDM is a numerical method for solving differential equations by approximating them with difference equations, in which finite differences approximate the derivatives. FDMs are thus discretization methods. FDMs convert a linear (non-linear) ODE (Ordinary Differential Equations)/PDE (Partial differential equations) into a system of linear (non-linear) equations, which can then be solved by matrix algebra techniques. The reduction of the differential equation to a system of algebraic equations makes the problem of finding the solution to a given ODE ideally suited to modern computers, hence the widespread use of FDMs in modern numerical analysis. The finite difference method is also one of a family of methods for approximating the solution of partial differential equations such as heat transfer, stress/strain mechanics problems, fluid dynamics problems, electromagnetics problems, etc. [34].

By discretizing the domain into grid of spaced points, the first derivative and the second derivative are approximated using the following formulas:

$$\frac{du}{dx} = \frac{u_{n+1} - u_n}{\Delta x} \quad \text{Forward Euler Method} \tag{2.2}$$

$$\frac{du}{dx} = \frac{u_n - u_{n-1}}{\Delta x} \quad \text{Backward Euler Method} \tag{2.3}$$

$$\frac{d^2 u}{dx^2} = \frac{u_{n-1} - 2u_n + u_{n+1}}{\Delta x^2} \tag{2.4}$$

where Δx is the **step size**, after that solve the resultants system of linear or nonlinear equations.

2. Finite Element Method (Discretization)

Arbitrary shaped boundaries are difficult to implement in finite difference methods. So, finite element method (FEM) is used. FEM covers the space with finite elements. The elements do not need to have the same size and shape. We get function value and derivative by interpolation. FEM is a numerical method that is used to solve boundary value problems defined by a partial differential equation (PDE) and a set of boundary conditions. The first step of using FEM to solve a PDE is discretizing the computational domain into finite elements. Then, the PDE is rewritten in a weak formulation. After that, proper finite element spaces are chosen and the finite element scheme is formed from the weak formulation. The next step is calculating those element matrices on each element and assembling the element matrices to form a global linear system. Then, the boundary conditions are applied, the sparse linear system is solved, and finally, post-processing of the numerical solution is done [35, 36]. An example to describe heat equation and show the boundary conditions is shown in Fig. 2.5.

3. Boundary Element Method (BEM)

BEM is the method of choice for applications requiring analysis of space around a device, and the exact modeling of boundaries. BEM uses an **integral** formulation of Maxwell's Equations, which allows for very accurate field calculations. Unlike FEM the electric and magnetic fields are computed directly from the source. This technique produces accuracies not attainable by Finite Element Method. Therefore, the basic difference between these two techniques is the fact that BEM only needs to solve for unknowns on the boundaries, whereas FEM solves for unknowns in the volume. Thanks to BEM, only active regions

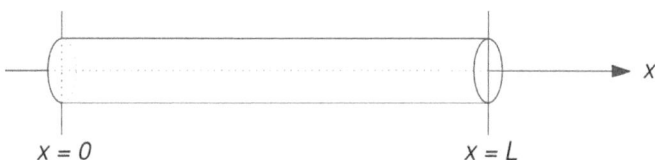

$$X = 0 \qquad\qquad X = L$$

PDE: $\alpha^2 u_{xx} = u_t$ for $0 \le x \le L$ and $t > 0$

Boundary values: $u(0,t) = 0,$ $u(L,t) = 0$ for $t > 0$

Initial values: $u(x,0) = f(x)$ for $0 \le x \le L.$

$u(x,t) =$ Temperature at point x and time t

$\alpha^2 =$ thermal diffusivity constant

Fig. 2.5 An example to describe heat equation

require discretization, allowing fields to be calculated anywhere in the "world". The combination of these two solvers offers exceptional facility in the analysis of electromagnetic problems.

Method of momentum (**MoM**) is another naming convention for BEM. MoM is usually a surface meshing approach. Volume meshing is more appropriate for most 3-D arbitrary geometries. MoM is often specialized to planar geometries and is more appropriate for most planar geometries.

4. Finite Differences Time Domain (FDTD)

FDTD is the application of finite differences to Maxwell's equations, in a second order, stable, staggered-grid approach for electric and magnetic fields. Some of the main advantages of the FDTD is that being based on the time domain, the FDTD method supports a wide range of frequencies, and additionally, the incorporation of non-linear materials is straightforward. The flexibility of time-domain discretization comes at a certain price, tough. As with any finite difference method, the propagation of a wave in the discrete grid doesn't obey the exact dispersion relations of Maxwell's equations, but rather an approximate version of them. This is called numerical dispersion error, and it can quickly become one of the main accuracy limitations of the FDTD. FDTD is a Time Domain method, meaning that it iterates over time [37, 38].

5. Partial Element Equivalent Circuit (PEEC)

The PEEC approach is a full wave electromagnetic electrical modeling technique for conductors embedded in arbitrary dielectrics in terms of equivalent circuits. The basic formulation is an electric field integral equation (EFIE) full wave solution to Maxwell's equations. The models can be used in both the time as well as the frequency domain. It facilitates the solution of problems which have both an electromagnetic part as well as a circuit part. Also, it leads to an intuitive understanding of electromagnetic problems [39].

6. Legendre Polynomials

It is used for expanding functions and solving **certain types of differential** equations. Legendre Polynomials are solutions to Legendre's ODE which can be expressed as follows [40, 41].

$$\left(1 - x^2\right)\frac{dy^2}{d^2x} - 2x\frac{dy}{dx} + k(k+1)y = 0 \tag{2.5}$$

The notation for expressing Legendre Polynomials is P_k and examples of the first 7 Legendre Polynomials are:

$$P_0(x) = 1 \tag{2.6}$$

$$P_1(x) = x \tag{2.7}$$

$$P_2(x) = \frac{1}{2}(3x^2 - 1) \tag{2.8}$$

$$P_3(x) = \frac{1}{2}(5x^3 - 3x) \tag{2.9}$$

$$P_4(x) = \frac{1}{8}(35x^4 - 30x^2 + 3) \tag{2.10}$$

$$P_5(x) = \frac{1}{8}\left(63x^5 - 70x^3 + 15x\right) \tag{2.11}$$

$$P_6(x) = \frac{1}{16}(231x^6 - 315x^4 + 105x^2 - 5) \tag{2.12}$$

The solution to Legendre's ODE is a series solution and can be written as

$$y = \sum_{n=0}^{\infty} a_n x^n \tag{2.13}$$

The real question is how we can get the values of a_n. By following a long term derivation, we can reach a recurrence relation that shows that a value of a_n depends on the values 2 steps before it. Which means that if we have a_0 and a_1 we can derive the rest of the series.

$$a_{n+2} = \frac{a_n(n-k)(n+k+1)}{(n+2)(n+1)} \tag{2.14}$$

When we start expanding these coefficients, we notice that the resultant series can be written as 2 series, one with odd exponents and the other with even exponents. If k is odd then the odd series converges and the even series diverges. On the other hand, if k is even then the odd series diverges and the even series converges. The convergent series can then be used as the as the Legendre Polynomial P_k for this k. A general closed form for computing Legendre Polynomials has been derived and is as follows.

$$P_k(x) = \sum_{n=0}^{LOOP\_LIMIT} (-1)^n \frac{(2k-2n)!}{2^k n!(k-n)!(k-2n)!} x^{k-2n} \tag{2.15}$$

The parameter *LOOP_LIMIT* can be computed as follows:

$$LOOP\_LIMIT = \begin{cases} k/2, & if \ k \ is \ even \\ (k-1)/2, & if \ k \ is \ odd \end{cases} \tag{2.16}$$

An Example of Green's function in Circuit Theory is given by:

$$L\frac{di}{dt} + Ri = v(t)$$

7. Green's Functions

Green's function is a type of function used to solve **inhomogeneous differential equations** subject to boundary conditions. The exact form of the Green's function depends on the differential equation, the body shape, and the type of boundary conditions present [42].

8. Runge–Kutta Methods

Runge–Kutta methods constitute a family of numerical techniques employed for solving ordinary differential equations (ODEs). They operate by computing a weighted average of multiple derivative estimates at a given point to approximate the function's value at the subsequent point. The accuracy of the approximation increases with the order of the Runge–Kutta method utilized. However, employing higher-order methods demands greater computational resources. Hence, the selection of an appropriate Runge–Kutta method involves balancing the desired level of accuracy with computational efficiency [43].

2.3.2   Hybrid Methods for Solving PDEs and ODEs

In [44], the authors propose a novel hybrid analytical numerical method for solving certain linear PDEs. The new method has advantages in comparison with classical methods, such as avoiding the solution of ordinary differential equations that result from the classical transforms, as well as constructing integral solutions in the complex plane which converge exponentially fast and which are uniformly convergent at the boundaries.

2.3.3   ML-Based Methods for Solving ODEs and PDEs

Solving high-dimensional PDEs using traditional methods is extremely challenging due to complex mathematical expressions. NN can be a powerful deep learning method that incorporates physical principles into the neural network algorithm, resulting in superior approximation and generalization capabilities. This has made it extremely popular in solving high-dimensional partial differential equations (PDEs).

The authors in [45] present a novel approach based on modified artificial **neural network** and optimization technique to solve partial differential equations. In [46], the authors propose using genetic algorithm to solve PDEs. Genetic algorithm is based on iterative procedures of search for an optimal solution for a problem which has multiple local minima or maxima. The algorithm passes through steps of recombination including crossover, and mutation then selection which increase the probability of finding the most optimum solution, which is the reduced model with the least error compared to the original transfer model. The error is compared with the original transfer function in terms of fitness function. Before applying the genetic operators, a method of encoding should be chosen to represent the data either in float form which is the raw form of the data or binary representation or any other representation. The crossover operator is a process of creating new individuals by selecting two or more parents and passing them through crossover procedures producing one or two or more individuals. Unlike real life there is no obligation to be abided by nature rules, so the new individual can have more than two parents. There is more than one method for crossover process like simple crossover which includes exchange of genes between the two chromosomes according to a specified crossover rate. Arithmetic crossover occurs by choosing two or more individuals randomly from the current generation and multiplying them by one random in the case of the Inevitability of the presence of a certain defined domain and more than one random if there is no physical need for a defined search domain.

The second process of genetic operators is mutation which is a process that occurs to prevent falling of all solutions of the population into a local optimum of the problem. The third genetic operator is selection which is the process of choosing a certain number of individuals for the next generation to be recombined generating new individuals aiming to find the most optimum solution. There is more than one technique to select individuals for the next generation. The Elitism selection is simply selecting the fittest individuals from the current population for the next population. This method guarantees a high probability of getting closer to the most optimum solution due to passing of the fittest chromosomes to the crossover operator producing fitter individuals. Another technique is Roulette wheel selection; in this kind of selection the parenthood probability is directly proportional to the fitness of the individuals where every individual is given a weight proportional to its fitness having a higher probability to be chosen as a parent for the next generation individuals and this technique is similar to rank selection where all individuals in the current population are ranked according to their fitness. Each individual is assigned a weight inversely proportional to the rank. The fitness function is the common evaluation factor among all selection techniques. Fitness function is a function which evaluates the fitness of each individual to be selected for the next generation [47]. The authors in [48–50] provide novel approaches for solving the optimal control problem for PDEs using Physics-Informed Neural Networks (PINNs).

2.3.4   How to Choose a Method for Solving PDEs and ODEs

When it comes to solving partial differential equations (PDEs) or ordinary differential equations (ODEs), there are various methods to choose from, depending on the specific problem at hand. Here are some factors to consider when choosing a method:

- Type of equation: The type of equation you're trying to solve (e.g. linear, nonlinear, first order, second order, etc.) will determine which methods are applicable.
- Boundary conditions: The boundary conditions of the problem can also dictate which methods are appropriate. For example, if the problem has periodic boundary conditions, Fourier series methods may be useful.
- Accuracy requirements: Depending on the level of accuracy required, some methods may be more suitable than others. For example, high-order finite difference methods may be necessary for very accurate solutions.
- Computational resources: Some methods require more computational resources (e.g. memory, CPU time) than others. Depending on the available resources, you may need to choose a method that is less computationally intensive.
- Numerical stability: Some methods are more numerically stable than others, which means they are less prone to generating numerical errors or diverging. This is an important consideration when choosing a method.
- Complexity of implementation: Some methods are easier to implement than others. Depending on your level of expertise and the available resources, you may need to choose a method that is relatively easy to implement.

As stated earlier, there exists a variety of algorithms for solving PDEs and ODEs (Fig. 2.6). Selecting the best solver algorithm depends on the number of iterations and memory usage. Comparison between different methods in terms of computation time and number of iterations is shown in Table 2.4. The criteria on which we can choose which method to be used is summarized in Table 2.5.

Fig. 2.6 Different methods of solving PDEs and ODEs

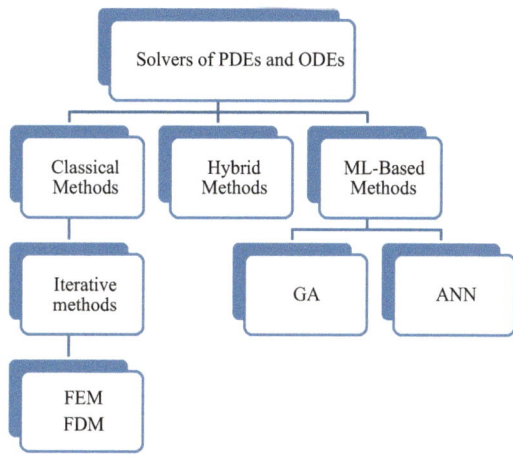

Table 2.4 Comparison between different iterative methods

Methods	Computation time (s)	Number of iterations	Memory usage
FEM	7	100	High
FDM	9	170	High

Table 2.5 How to choose a method for solving nonlinear equations

PDEs and ODEs solver methods	Boundaries	
	Arbitrary shaped	Rectangular
FEM	✓	✓
FDM	✗	✓

2.4   Different Methods for Solving System of Non-linear Equations (SNLEs)

The nonlinear systems problems are important in physics and engineering domains. For example, device simulation involves solving a large system of non-linear equations. So, most simulation time is spent for solving this huge number of equations.

A nonlinear system of equations is a system in which at least one of the equations is nonlinear. For example, a system that contains one quadratic equation and one linear equation is a nonlinear system. A system made up of a linear equation and a quadratic equation can have no solution, one solution, or two solutions, as shown in Fig. 2.7. Nonlinear systems can be solved **analytically by graphing, substitution, or eliminations.**

 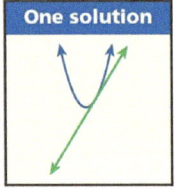 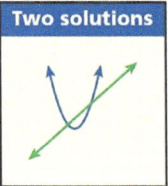

Fig. 2.7 Solution of linear and nonlinear equations

The substitution method is a good choice when equation is solved for a variable, both equations are solved for the same variable, or a variable in either equation has a coefficient of 1 or − 1.

The elimination method is a good choice when both equations have the same variable term with the same or opposite coefficients or when a variable term in one equation is a multiple of the corresponding variable term in the other equation. In this work, we focus on the numerical methods not the analytical methods.

A lot of techniques that are used for nonlinear systems come from linear systems, because nonlinear systems can sometime be approximated by linear systems.

A nonlinear equation involves terms of degree higher than one. Any nonlinear equation can be expressed as

$$f(x) = 0.$$

They cannot be solved using direct methods, but using iterative methods. In this section, all methods to solve SNLEs are discussed. The explanation of the methods will be on 3 variables nonlinear equations shown below:

$$3x_1 - \cos x_2 x_3 - 0.5 = 0$$

$$x_1^2 - 81(x_2 + 0, 1)^2 + \sin x_3 + 1.06 = 0$$

$$e^{-x_2 x_1} + 20x_3 + \frac{10\Pi - 3}{3} = 0 \tag{2.17}$$

2.4.1 Iterative Methods for Solving SNLEs

1. *Newton Method and Newton-Raphson Method*

Nonlinear problems are often treated numerically by reducing them to a sequence of linear problems. As a simple but important example, consider the problem of solving a

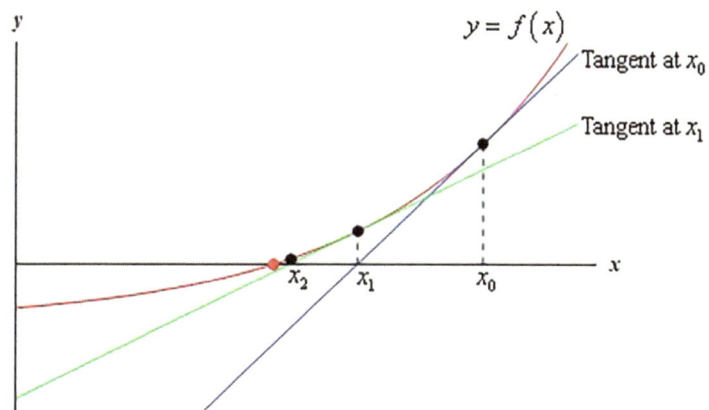

Fig. 2.8 Newton method

nonlinear equation f(x) = 0. Approximate the graph of y = f(x) by the tangent line at a point x(0) near the desired root and use the root of the tangent line to approximate the root of the original nonlinear function f(x). This leads to Newton's iterative method for finding successively better approximations to the desired root as shown in Fig. 2.8.

This method originates from the **Taylor's series** expansion of the function *f(x)* about the point x_1 as follows [51]:

$$f(x) = f(x_1) + (x - x_1)f'(x_1) + \frac{1}{2!}(x - x_1)^2 f''(x_1) + \cdots \approx f(x_1) + (x - x_1)f'(x_1) = 0$$

$$(2.18)$$

Rearrange (2.18) results in (2.19):

$$x = x_1 - \frac{f(x_1)}{f'(x_1)}$$

$$(2.19)$$

Generalizing (2.19) we obtain Newton's iterative method that can only be used to solve nonlinear equations involving only a single variable:

$$x_{n+1} = x_n - \frac{f(x_n)}{f'(x_n)}$$

$$(2.20)$$

To be able to use (2.20) in solving nonlinear equations involving many variables, we define the Jacobian matrix is a matrix of first order partial derivatives. This is called Newton–Raphson Method. So, Newton method is used to solve single variable nonlinear equations, while Newton–Raphson Method is used to solve multivariable nonlinear equations [52].

$$J(x) = \begin{bmatrix} \frac{df_1}{dx_1} & \frac{df_1}{dx_2} & \frac{df_1}{dx_3} \\ \frac{df_2}{dx_1} & \frac{df_2}{dx_2} & \frac{df_2}{dx_2} \\ \frac{df_3}{dx_1} & \frac{df_3}{dx_2} & \frac{df_3}{dx_2} \end{bmatrix} \tag{2.21}$$

So, the following iterative equation is used for solving nonlinear equations involving many variables.

$$x_{n+1} = x_n - J(x_0)^{-1} f(x_n) \tag{2.22}$$

The detailed steps of the solution of Eq. (2.17) can be shown as follows.

- Step 1: Let $x_1 = x_2 = -x_3 = 0.1$
- Step 2: Find F (x)

$$F(x) = \begin{bmatrix} 3x_1 - \cos x_2 x_3 - 0.5 \\ x_1^2 - 81(x_2 + 0, 1)^2 + \sin x_3 + 1.06 \\ e^{-x_2 x_1} + 20x_3 + \frac{10\Pi - 3}{3} \end{bmatrix} \tag{2.23}$$

- Step 3: Find the Jacobi Matrix

$$J(x) = \begin{bmatrix} 3 & x_3 \sin x_2 x_3 & x_2 \sin x_2 x_3 \\ 2x_1 & -162(x_2 + 0.1) & \cos x_3 \\ -x_2 e^{-x_2 x_1} & -x_1 e^{-x_2 x_1} & 20 \end{bmatrix} \tag{2.24}$$

- Step 4: Find F (x_0) and J (x_0)

$$F(x_0) = \begin{bmatrix} 0.3 - \cos -0.01 - 0.5 \\ 0.01 - 3.24 + \sin -0.1 + 1.06 \\ e^{-0.01} - 2 + \frac{10\Pi - 3}{3} \end{bmatrix} \tag{2.25}$$

$$J(x_0) = \begin{bmatrix} 3 & -0.1 \sin -0.01 & 0.1 \sin -0.01 \\ 0.2 & -32.4) & \cos -0.1 \\ -0.1e^{-0.01} & -0.1e^{-0.01} & 20 \end{bmatrix} \tag{2.26}$$

- Step 5: Apply into the following equation:

$$x_1 = x_0 - J(x_0)^{-1} f(x_0) \tag{2.27}$$

The solution of this step is $\begin{bmatrix} 0.5 \\ 0.01 \\ -0.5 \end{bmatrix}$.

- Step 6: Use the results of x_1 to find our next iteration x_2 by using the same procedure.

The final solution is $\begin{bmatrix} 0.5 \\ 0 \\ -0.5 \end{bmatrix}$.

One of the advantages of Newton's method is its simplicity. The major disadvantage associated with Newton's method, is that $J(x)$, as well as its inversion has, to be calculated for each iteration which consumes time depending on the size of your system is. The convergences of the classical solvers such as Newton-type methods are highly sensitive to the initial guess of the solution. So, it might not converge if the initial guess is poor. Newton method flowchart is shown in Fig. 2.9. Newton–Raphson is used to find the reciprocal of D and multiply that reciprocal by N to find the final quotient Q.

2. Quasi-Newton method *a.k.a* Broyden's method

It uses an approximation matrix that is updated at each iteration instead of the Jacobian matrix. So, the iterative procedure for Broyden's method is the same as Newton's method except that an approximation Matrix A is used instead of Jacobi method. The main advantage of Broyden's method is the reduction of computations, but it needs more iterations than Newton's method [53, 54]. Quasi-Newton method is a method used to either find zeroes or local maxima and minima of functions, as an alternative to Newton's method. As the most important disadvantage of Newton's method is the requirement that $F'(xk)$ be determined for each k (number of iteration), this is a very costly operation, but the exact cost varies from problem to problem based on its complexity. It works as follows:

- Step 1: Given a starting point $\mathbf{x}\,(0) \in \mathbf{Rn}$, $\mathbf{H(0)} > 0$. Where $\mathbf{H}$ is the Hessian matrix of the equation, $\mathbf{H^{-1}}$ is the inverse of the Heassian matrix and $\mathbf{t}$ is the step size. H is given by Eq. (1)
- Step 2: For k (iteration number until reaching an acceptable error of x) = 1, 2, 3, ..., repeat:

    1. Compute Quasi-Newton direction:
       $$\Delta x(k-1) = -\nabla f(x(k-1) * H^{-1}(k-1)$$
    2. Update x (k) = x (k− 1) + tΔx (k− 1).
    3. Compute H(k).

Different methods use different rules for updating H in step 3. These methods are summarized in Table 2.6. Advantages of Quasi-Newton method:

- Computationally cheap compared with Newton's method.
- Faster computation.

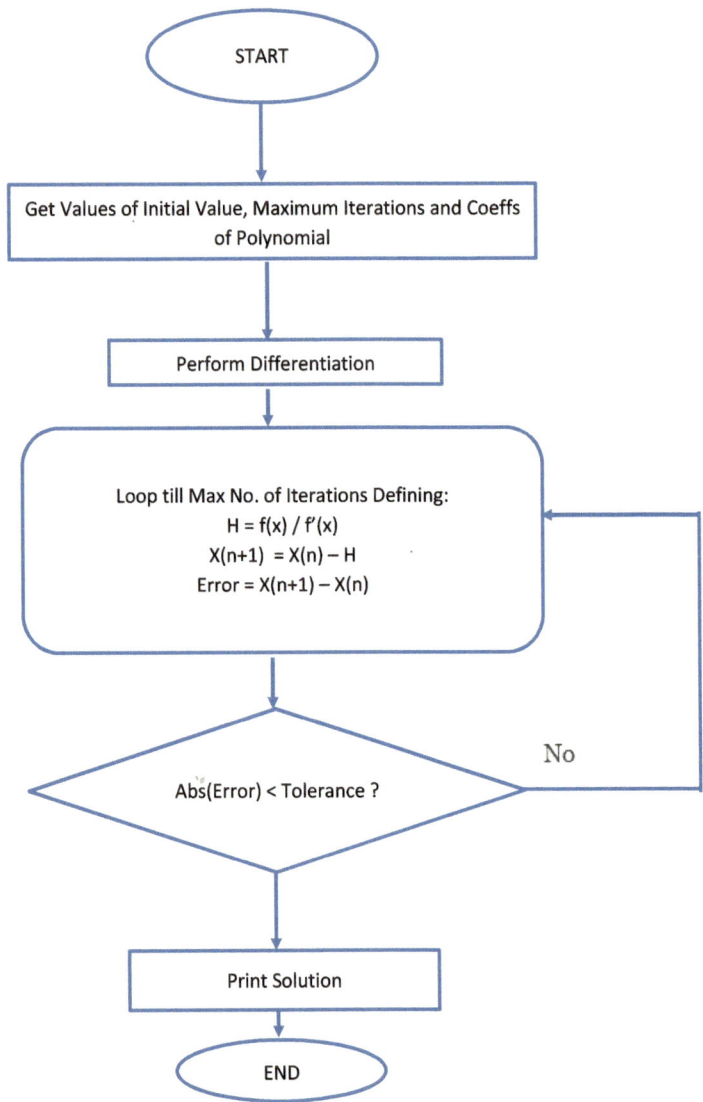

Fig. 2.9 Newton method flow-chart

- No need for second derivative.

Disadvantages:

- More convergence steps.
- Less precise convergence path.

Table 2.6 H updates methods

Update method	Equation
Broyden-Fletcher-Goldfarb-Shanno (BFGS) update	$H_k = H_{k-1} + \frac{yy^T}{y^T s} - \frac{H_{k-1}ss^T H_{k-1}}{s^T H_{k-1}s}$ $s = x^{(k)} - x^{(k-1)}, \quad y = \nabla f(x^{(k)}) - \nabla f(x^{(k-1)})$
Symmetric rank one update (B is the Heassian matrix)	$B^+ = B + \frac{(y-Bs)(y-Bs)^T}{(y-Bs)^T s}$
Davidon-Fletcher-Powell (DFP) update	$H_k = \left(I - \frac{ys^T}{s^T y}\right)H_{k-1}\left(I - \frac{sy^T}{s^T y}\right) + \frac{yy^T}{s^T y}$

$$
H = \begin{bmatrix}
\frac{\partial^2 f}{\partial x_1^2} & \frac{\partial^2 f}{\partial x_1 x_2} & \cdots & \frac{\partial^2 f}{\partial x_1 x_n} \\
\frac{\partial^2 f}{\partial x_2 x_1} & \frac{\partial^2 f}{\partial x_2^2} & \cdots & \frac{\partial^2 f}{\partial x_2 x_n} \\
\vdots & \vdots & \ddots & \vdots \\
\frac{\partial^2 f}{\partial x_n x_1} & \frac{\partial^2 f}{\partial x_n x_2} & \cdots & \frac{\partial^2 f}{\partial x_n^2}
\end{bmatrix} \tag{2.28}
$$

3. The Secant Method

In this method, we replace the derivative $f'(x_i)$ in the Newton–Raphson method with the following linearization equation [55, 56]:

$$
f'(x_{n+1}) = \frac{f(x_n) - f(x_{n+1})}{x_n - x_{n+1}} \tag{2.29}
$$

Moreover, in contrast to Newton Raphson method, here we need to provide two initial values of x to get the algorithm started (x_0, x_1). In other words, we approximate the function by a straight line (Figs. 2.10 and 2.11).

The absolute error in every iteration can be calculated as

$$
e = |x_n - x_{n-1}| \tag{2.30}
$$

Iteration is stopped if one of the following conditions is met:

- Number of iterations exceeds a certain number
- Error is below a certain threshold

However, the secant method is not always guaranteed to converge. There is no guarantee that the secant method will converge to the root in the following scenarios:

Fig. 2.10 Secant line for the
first two guesses

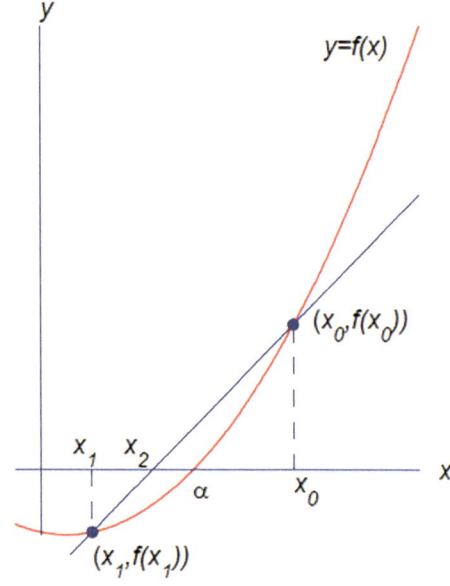

Fig. 2.11 Secant line for the
new estimate with the second
guess from previous iteration

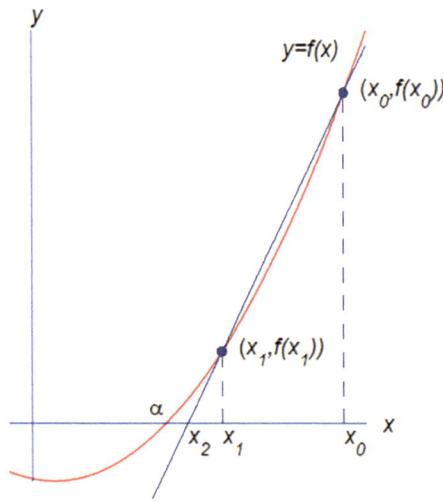

- If the initial estimates are not close enough to the root.
- If the value of the first derivative of the function at the root equals zero.

Advantages:

- Does not require the calculation of a derivative, unlike Newton Raphson.

- Relatively fast rate of convergence (faster than linear, but slower than the Newton Raphson method)

Disadvantages:

- Requires two initial guesses not one like the Newton Raphson Method.
- May not converge (if the conditions mentioned above are not met).

4. The Muller Method

It is based on Secant method, but instead of linear approximation, it uses quadratic approximate. It works only for a set of polynomial equations. It begins with three initial assumptions of the root, and then constructing a parabola through these three points, and takes the intersection of the x-axis with the parabola to be the next approximation. This process continues until a root with the desired level of accuracy is found [57].

2.4.2 Hybrid Methods for Solving SNLEs

The iterative methods are categorized as either locally convergent or globally convergent. The locally convergent methods have fast convergence rate, but they require Initial approximations close to the roots. While the globally convergent methods have very slow convergence rate and Initial approximations far from the roots should lead to convergence.

Hybrid methods combine locally and globally convergent methods in the same algorithm such that if the initial approximations are far from the roots, the slow globally convergent algorithm is used until the points are close enough to the roots then the locally convergent method is used.

In [58], the authors propose a hybrid method for solving systems of nonlinear equations that require fewer computations than those using regular methods. The proposed method presents a new technique to approximate the Jacobian matrix needed in the process, which still retains the good features of Newton's type methods, but which can reduce the run time.

In [59], authors propose a hybrid approach for solving systems of nonlinear equations. The approach is based on using chaos optimization technique with quasi-newton method to speed-up run time.

2.4.3 ML-Based Methods for Solving SNLEs

The authors in [60] present a novel framework for the numerical solution of nonlinear differential equations using neural networks with back propagation algorithm.

In [16], the authors propose using genetic algorithm to solve SNLEs. GA approach optimizes the space and time complexities in solving the nonlinear system as compared to the traditional computing methods.

2.4.4   How to Choose a Method for Solving Non-linear Equations

As stated earlier, there exists a variety of algorithms for solving Non-linear systems (Fig. 2.12). Selecting the best solver algorithm depends on the number of iterations and memory usage.

Comparison between different methods based on Eq. (2.31) in terms of computation time and number of iterations is shown in Table 2.7.

One of the advantages of Newton's method is its simplicity. The major disadvantage associated with Newton's method, is that J(x), as well as its inversion, has, to be calculated for each iteration which consume time depending on the size of your system is. The main advantage of Broyden's method is the reduction of computations, but it needs more iterations than Newton's method. The criteria on which we can choose which method to be used are summarized in Table 2.8.

Fig. 2.12 Different methods of solving systems of non-linear equations

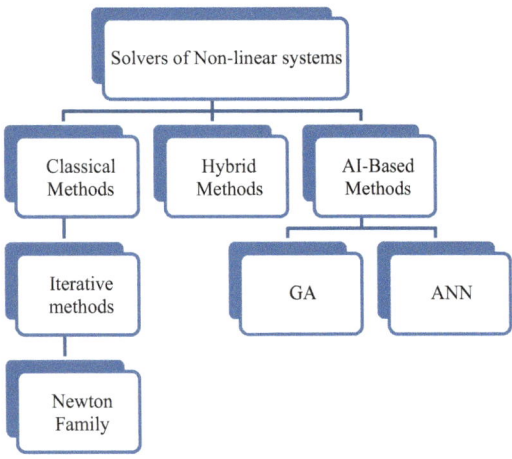

Table 2.7 Comparison between different iterative methods

Methods	Computation time (s)	Number of iterations	Memory usage
Newton	3	13	The highest
Quasi-Newton	5	20	Less
Secant	4	10	Less
Muller	4	9	Less

Table 2.8 How to choose a method for solving nonlinear equations

SNLEs solver methods	Jacobian matrix type		Converge
	Sparse	Dense	
Newton	✓	✗	Locally
Quasi-Newton	✗	✓	Locally
Secant	✗	✓	Globally
Muller	✗	✓	Globally

2.5   Different Methods for Solving System of Linear Equations (SLEs)

Linear algebraic primitives are at the core of many modern algorithms in engineering, science, physics, and machine learning. For example, circuit simulation involves solving a large system of non-linear equations. So, most simulation time is spent for solving this huge number of equations. A linear system of equations can be represented as $Ax = b$.

It is important to consider that not all matrices can be solved using iterative methods. A matrix can be solved only when it is diagonally dominant. A matrix is said to be diagonally dominant if the magnitude of every diagonal entry is more than the sum of the magnitude of all the non-zero elements of the corresponding row. Both methods sometimes converge even if this condition is not satisfied. However, it is necessary that the magnitude of diagonal terms in a matrix is greater than the magnitude of other terms. A determinant equal zero means that the matrix is singular and the system is ill-conditioned [15]. In this section, all methods to solve SLEs are discussed. The explanation of the methods will be on 3×3 linear equations shown below:

$$\begin{bmatrix} 4 & 2 & 3 \\ 3 & -5 & 2 \\ -2 & 3 & 8 \end{bmatrix} \begin{bmatrix} x \\ y \\ z \end{bmatrix} = \begin{bmatrix} 8 \\ -14 \\ 27 \end{bmatrix} \tag{2.31}$$

2.5.1   Direct Methods for Solving SLEs

1. Cramer's Rule Method

In linear algebra, Cramer's rule is an explicit formula for the solution of systems of linear equations with as many equations as unknowns; it is valid only when the system has a unique solution. It expresses the solution in terms of the determinants of the square coefficient matrix and of matrices obtained from it by replacing one column by the vector of right hand sides of the equations [51].

The value of the determinant T should be checked because if $T = 0$, the equations wouldn't have a unique solution. The main advantage of this design is that it is very fast compared to other methods as it needs fewer number of clock cycles to calculate the result. The main disadvantage of this design is that it consumes a lot of resources. As a result, it is not a scalable solver. Cramer method is preferable to be used with square matrix. It relies on finding the inverse of matrix A. So, the solution to $Ax = b$ is $x = A^{-1}b$ [61–63].

The detailed steps of the solution of Eq. (2.17) can be shown as follows.

- Step 1: Write down the main matrix and find its determinant Δ.
- Step 2: Replace the 1st column of the main matrix with the solution vector (b) and find its determinant Δ_1.
- Step 3: Replace the 2nd column of the main matrix with the solution vector and find its determinant Δ_2.
- Step 4: Replace the 3rd column of the main matrix with the solution vector and find its determinant Δ_3.
- Step 5: Calculate the output using the following formulas $x = \frac{\Delta_1}{\Delta}$; $y = \frac{\Delta_2}{\Delta}$, $z = \frac{\Delta_3}{\Delta}$.
- Step 6: Solution $= \begin{bmatrix} -1 \\ 3 \\ 2 \end{bmatrix}$.

Cramer's rule is computationally inefficient for systems of more than two or three equations. In the case of n equations in n unknowns, it requires computation of n + 1 determinants, while Gaussian elimination produces the result with the same computational complexity as the computation of a single determinant. Cramer's rule can also be numerically unstable even for 2×2 systems. However, it has recently been shown that Cramer's rule can be implemented in $O(n^3)$ time, which is comparable to more common methods of solving systems of linear equations, such as Gaussian elimination (consistently requiring 2.5 times as many arithmetic operations for all matrix sizes), while exhibiting comparable numeric stability in most cases. In linear algebra Cramer's rule is an explicit formula for the solution of a system of linear equation. Cramer's rule flowchart is shown in Fig. 2.13.

The determinant is calculated in a recursive function, that reduces the input matrix to multiple smaller matrices and calls itself with each of these matrices, until the size of these smaller matrices becomes 2×2, then it computes its determinant and returns it to the previous function call, which uses these values to obtain the result of the original input matrix. Determinant Calculation example is shown in Fig. 2.14.

2. Gaussian Elimination Method

Gaussian elimination (also known as row reduction) is an algorithm for solving systems of linear equations. It is usually understood as a sequence of operations performed on

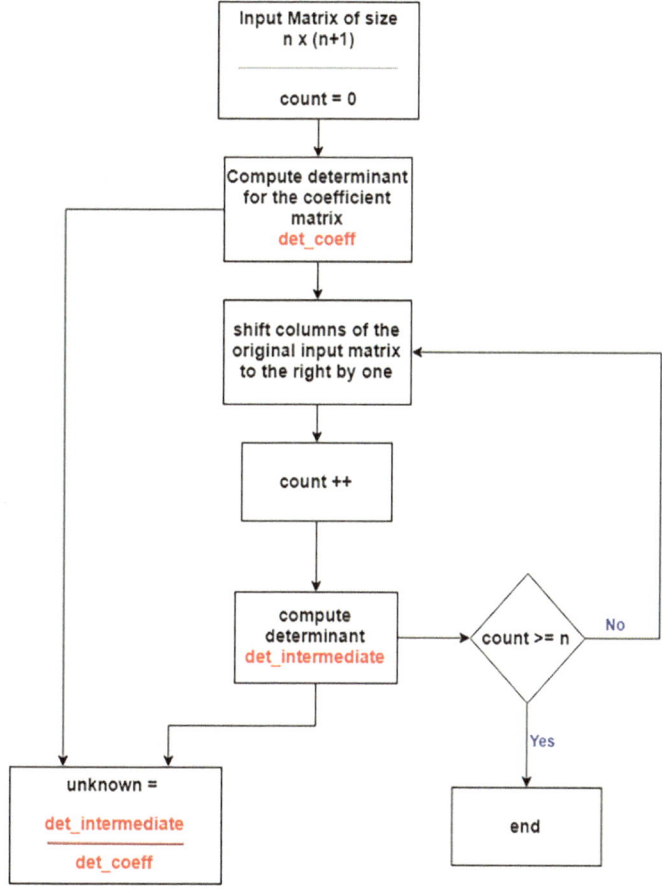

Fig. 2.13 Cramer's method flowchart

the corresponding matrix of coefficients. This method can also be used to find the rank of a matrix, to calculate the determinant of a matrix, and to calculate the inverse of an invertible square matrix. To perform row reduction on a matrix, one uses a sequence of elementary row operations to modify the matrix until the lower left-hand corner of the matrix is filled with zeros, as much as possible. There are three types of elementary row operations [64, 65]:

– Swapping two rows,
– Multiplying a row by a nonzero number,
– Adding a multiple of one row to another row.

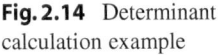
Fig. 2.14 Determinant
calculation example

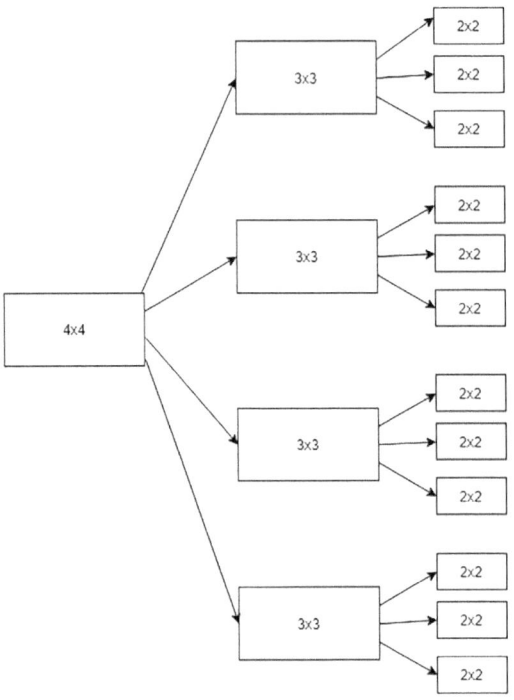

Using these operations, a matrix can always be transformed into an upper triangular matrix, and in fact one that is in row echelon form. Once all of the leading coefficients (the leftmost nonzero entry in each row) are 1, and every column containing a leading coefficient has zeros elsewhere, the matrix is said to be in reduced row echelon form. This final form is unique; in other words, it is independent of the sequence of row operations used.

In gauss elimination, forward elimination is used first to eliminate all the elements in the lower triangle of the matrix. Next, backward elimination is used to eliminate all the elements in the upper triangle of the matrix, leaving only the main diagonal. The main disadvantage of this design is that it consumes a lot of logic resources, but it is fast [45, 46]. The detailed steps of the solution of Eq. (2.1) can be shown as follows [61, 62].

- Step 1: Make the pivot in the 1st column. In other words, obtain row echelon form for the matrix.
- Step 2: Make the pivot in the 2nd column.
- Step 3: Make the pivot in the 3rd column.
- Step 4: The output can be found from the matrix which has all ones as diagonal elements and zeros in lower triangles.

- Step 5: Solution $= \begin{bmatrix} -1 \\ 3 \\ 2 \end{bmatrix}$.

3. Gauss Jordan (GJ) Elimination Method

GJ elimination is a method based on Gaussian elimination that puts zeros above and below each pivot giving in this way the inverse matrix. The differences from Gaussian elimination method are: When an unknown is eliminated from an equation, it is also eliminated from all other equation and all rows are normalized by dividing them by their pivot element. Hence, the elimination step results in an identity matrix rather than a triangular matrix. Therefore, back substitution is not necessary [63, 66]. The detailed steps of the solution of Eq. (2.27) can be shown as follows.

- Step 1: Make the pivot in the 1st column then eliminate it. In other words, obtain reduced row echelon form for the matrix, where the lower left part of this matrix contains only zeros and all of the zero rows are below the non-zero rows. The matrix is reduced to this form by the elementary row operations: swap two rows, multiply a row by a constant, add to one row a scalar multiple of another.
- Step 2: Make the pivot in the 2nd column then eliminate it.
- Step 3: Make the pivot in the 3rd column then eliminate it.
- Step 4: The output can be found from the unity matrix
- Step 5: Solution $= \begin{bmatrix} -1 \\ 3 \\ 2 \end{bmatrix}$.

Row reduction is the process of performing row operations to transform any matrix into (reduced) row echelon form. In reduced row echelon form, each successive row of the matrix has fewer dependencies than the previous, so solving systems of equations is a much easier task. The idea behind row reduction is to convert the matrix into an "equivalent" version in order to simplify certain matrix computations. Its two main purposes are to solve system of linear equations and calculate the inverse of a matrix. Carl Friedrich Gauss championed the use of row reduction, to the extent that it is commonly called **Gaussian elimination**. It was further popularized by Wilhelm Jordan, who attached his name to the process by which row reduction is used to compute matrix inverses, **Gauss-Jordan elimination**. A system of equations can be represented in a couple of different matrix forms. One way is to realize the system as the matrix multiplication of the coefficients in the system and the column vector of its variables. The square matrix is called the **coefficient matrix** because it consists of the coefficients of the variables in the system of equations. An alternate representation called an **augmented matrix** is created by stitching the columns of matrices together and divided by a vertical bar. The coefficient matrix is placed on the left of this vertical bar, while the constants on the right-hand side of each equation are placed on the right of the vertical bar. The matrices that represent these systems can be manipulated in such a way as to provide easy-to-read solutions. This manipulation is called row reduction. Row reduction techniques transform the matrix into **reduced row echelon form** without changing the solutions to the system. To convert any matrix to its reduced row echelon form, Gauss-Jordan elimination is performed. There are three elementary row operations used to achieve reduced row echelon form [67, 68]:

- Switch two rows.
- Multiply a row by any non-zero constant.
- Add a scalar multiple of one row to any other row.

4. LU decomposition Method

It is a form of Gaussian elimination method. It decomposes the A matrix into the form a lower triangular matrix and an upper triangular matrix. LU factorization is a computation intensive method, especially for large matrices [69]. It calculate $A = LU$, L is a lower triangular matrix with unit diagonal entries, and U is an upper triangular matrix. The detailed steps of the solution of Eq. (2.28) can be shown as follows.

- Step 1: let $LUx = b$.
- Step 2: let $Ux = y$.
- Step 3: Solve $Ly = b$ for y.
- Step 3: calculate x from step 2.
- Step 4: Solution $= \begin{bmatrix} -1 \\ 3 \\ 2 \end{bmatrix}$.

Another nice feature of the LU decomposition is that it can be done by overwriting A, therefore saving memory if the matrix A is very large. The LU decomposition is useful when one needs to solve Ax = b for x when A is fixed and there are many different b's. First one determines L and U using Gaussian elimination. Then one writes: (LU)x = L(Ux) = b. Then we let: y = Ux, and first solve Ly = b for y by forward substitution. We then solve Ux = y.

5. Cholesky decomposition Method

It is a decomposition of a Hermitian, positive-definite matrix into the product of a lower triangular matrix and its conjugate transpose, which is useful for numerical solutions [70, 71]. It is valid only for square and symmetric matrices only. So, it cannot be applied to our example in Eq. (2.32).

Cholesky decomposition is a special version of LU decomposition tailored to handle symmetric matrices more efficiently. For a positive symmetric matrix A where $a_{ij} = a_{ji}$, LU decomposition is not efficient enough as computational load can be halved using Cholesky decomposition. In Cholesky, matrix A can be decomposed as follows

$$A = L * L^T \tag{2.32}$$

where L is a low triangle matrix.
  To compute L:

$$l_{j,j} = \sqrt{a_{j,j} - \sum_{k=0}^{j-1} l_{j,k}^2} \tag{2.33}$$

$$l_{i,j} = \frac{1}{l_{j,j}} \left(a_{i,j} - \sum_{k=0}^{j-1} l_{i,k} l_{j,k} \right) \quad for \; i > j \tag{2.34}$$

2.5.2 Iterative Methods for Solving SLEs

Iterative methods for solving SLEs can be classified into two categories stationary methods such as Jacobi, Gauss–Seidel, and Gaussian over relaxation and Krylov subspace methods such as Conjugate gradient methods which extract the approximate solution from a subspace of dimension much smaller than the size of the coefficient matrix A. This approach is called projection method [53, 54, 61].
  The solution of the linear system Ax = b using stationary methods can be expressed in the general form.

$$MX^{n+1} = NX^n + b \tag{2.35}$$

where $x(k)$ is the approximate solution at the kth iteration. The choice of M and N for different stationary methods are summarized in Table 2.9.

Table 2.9 Stationary iterative methods for linear systems

Method	M	N
Jacobi	D	L + U
Gauss–Seidel	D − L	U
Gaussian over relaxation	$\frac{1}{\omega}D - L$	$\left(\frac{1}{\omega} - 1 \right)D + U$

D, L and U are the diagonal, lower-triangular and upper-triangular parts of A, respectively

For projection methods, let x_0 an initial guess for this linear system and $r_0 = b - Ax_0$ be its corresponding residual. We continue the iterations until r_0 is very small.

All these methods are discussed in details in the below subsections.

1. Jacobi Method

Jacobi as an iterative method for solving the system of linear equations. The Jacobi method starts with an initial guess of vector x and solves each unknown x. The obtained guess is then used as the current solution and the process is iterated again. This process continues until it converges to a solution [72].

To solve Eq. (2.32) using Jacobi method:

- Step 1: is to arrange Eq. (2.32) as below in an iterative format [73]:.

$$x = \frac{8 - 2y - 3z}{4} \tag{2.36}$$

$$y = \frac{-14 - 3x - 2z}{-5} \tag{2.37}$$

$$z = \frac{27 + 2x - 3y}{8} \tag{2.38}$$

- Step 2: substitute in the following equations assuming that x_0, y_0, z_0 are zeros.

$$x_{n+1} = \frac{8 - 2y_n - 3z_n}{4} \tag{2.39}$$

$$y_{n+1} = \frac{-14 - 3x_n - 2z_n}{-5} \tag{2.40}$$

$$z_{n+1} = \frac{27 + 2x_n - 3y_n}{8} \tag{2.41}$$

- Step 3: we continue in iterations until convergence happens. Solution $= \begin{bmatrix} -0.99 \\ 3 \\ 1.99 \end{bmatrix}$ after 20 iterations.

2. Gauss Seidel method

The Gauss–Seidel (GS) method, also known as the Liebmann method or the method of successive displacement, is an iterative method used to solve a linear system of equations. It is named after the German mathematicians Carl Friedrich Gauss and Philipp Ludwig von Seidel, and is similar to the Jacobi method. Though it can be applied to any matrix with non-zero elements on the diagonals, convergence is only guaranteed if the matrix is either diagonally dominant, or symmetric and positive definite. It was only mentioned in a private letter from Gauss to his student Gerling in 1823. A publication was not delivered before 1874 by Seidel.

In certain cases, such as when a system of equations is large, iterative methods of solving equations are more advantageous. Elimination methods, such as Gaussian elimination, are prone to large round-off errors for a large set of equations. Iterative methods, such as the Gauss–Seidel method, give the user control of the round-off error. Also, if the physics of the problem are well known, initial guesses needed in iterative methods can be made more judiciously leading to faster convergence.

GS is an improvement of the Jacobi algorithm. We use 'new' variable values (subscript $= n + 1$) wherever possible: substitute in the following equations assuming that x_0, y_0, z_0 are zeros and continue in iterations until convergence happens [16, 58–60].

$$x_{n+1} = \frac{8 - 2y_n - 3z_n}{4} \tag{2.42}$$

$$y_{n+1} = \frac{-14 - 3x_{n+1} - 2z_n}{-5} \tag{2.43}$$

$$z_{n+1} = \frac{27 + 2x_{n+1} - 3y_{n+1}}{8} \tag{2.44}$$

$$\text{Solution} = \begin{bmatrix} -0.99 \\ 3 \\ 1.99 \end{bmatrix} \text{ after 11 iterations.}$$

3. Successive over-relaxation (SOR) Method

It combines the parallel nature of Jacobi iterative method used with higher convergence rate. It is an enhanced version of Gauss Seidel method. We use a relaxation variable ω where generally $1 < \omega < 2$. Notice that if $\omega = 1$ then this *is* the Gauss–Seidel Method. The following equations assuming that x0, y0, z0 are zeros and continue in iterations until convergence happens.

$$x_{n+1} = \omega \frac{8 - 2y_n - 3z_n}{4} \tag{2.45}$$

$$y_{n+1} = \omega \frac{-14 - 3x_{n+1} - 2z_n}{-5} \qquad (2.46)$$

$$z_{n+1} = \omega \frac{27 + 2x_{n+1} - 3y_{n+1}}{8} \qquad (2.47)$$

Solution to Eq. (2.32) $= \begin{bmatrix} -0.99 \\ 3 \\ 1.99 \end{bmatrix}$ after 4 iterations.

It is often very difficult to estimate the optimal relaxation factor, which is a key parameter of the SOR method.

4. Conjugate Gradient method

Basic iterative methods such as Jacobi Method and Gauss Seidel method cannot solve all the linear systems. The Conjugate Gradient method is one of the Krylov subspace methods. The conjugate gradient (GC) method derives its name from the fact that it generates a sequence of conjugate (or orthogonal) vectors. These vectors are the residuals of the iterations. They are also the gradients of a quadratic functional, the minimization of which is equivalent to solving the linear system.

Conjugate Gradient is an Iterative Method applicable to sparse systems that are too large to be handled by a direct implementation. It has the advantage that it reaches the required tolerance after a relatively small number of iterations compared to Jacobi and gauss methods.

The conjugate gradient method is used to solve equations where the matrix is symmetric [72]. To solve Eq. (2.31) keep it in the form of $Ax = b$ then apply the following algorithm. It is valid only for square and symmetric matrices only. So, it cannot be applied to our example in Eq. (17).

The algorithm works as follows:

$$r_0 = b - Ax_0$$

$$p_0 = r_0$$

$$k = 0$$

Repeat

$$\alpha_n = \frac{r_n^T r_n}{p_n^T A p_n}$$

$$x_{n+1} = x_n + \alpha_n p_n$$

$$r_{n+1} = r_n - \alpha_n A p_n$$

If r_{n+1} is very small then exit the loop else

$$p_{n+1} = r_{n+1} + \beta_n p_n$$

$$n = n + 1$$

End repeat.
The solution is x_{n+1}.
Data-flow diagram for Conjugate Gradient method is shown in Fig. 2.15.

5. Bi-Conjugate Gradient method

Bi-Conjugate Gradient method can solve any linear system. The Bi-conjugate Gradient method generates two CG-like sequences of vectors, one based on a system with the original coefficient matrix, and one on its transposal. Instead of orthogonalizing each sequence, they are made mutually orthogonal, or bi-orthogonal. This method, like CG, uses limited storage. It is useful when the matrix is non-symmetric and nonsingular. A singular matrix is a square matrix that has no inverse. A determinant equal zero means that the matrix is singular. It is valid only for square and symmetric matrices only. So, it cannot be applied to our example in Eq. (2.32).

6. Generalized Minimal Residual method

The generalized minimal residual method (**GMRES**) is an iterative method for the numerical solution of a non-symmetric system of linear equations. It is based on **Arnoldi's algorithm** (an eigenvalue algorithm). GMRES solves at every step a minimum

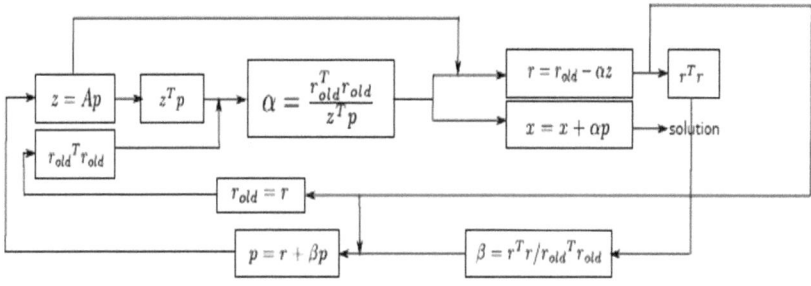

Fig. 2.15 Data-flow diagram for conjugate gradient method

squares problems (*min (Ax-b)*), which requires many computations. GMRES is useful if convergence is reached in a small number of iterations [74].

The generalized minimal residual method (GMRES) is an iterative method used to find numerical solutions to non-symmetric linear systems of equations. The method relies on constructing an orthonormal basis of the Krylov space and is thus vulnerable to an imperfect basis caused by computational errors. There have been attempts to address this issue by devising variations of the method that are less sensitive to poorly conditioned problems. The GMRES algorithm is typically used when the dimensions of the problem are very large, thus it is of interest to investigate ways in which the computational and memory cost of running it can be reduced.

- **Krylov subspace**: The order -r Krylov subspace generated by an n-by-n matrix A and a vector b of dimension n is the linear subspace spanned by the images of b under the first r powers of A (starting from $A^0 = I$), that is, $K_r(A, b) = span\{b, Ab, A^2b, \ldots, A^{r-1}b\}$. We use Krylov subspace in iterative methods for finding one (or a few) eigenvalues of large sparse matrices or solving large systems of linear equations avoid matrix–matrix operations, but rather multiply vectors by the matrix and work with the resulting vectors. Starting with a vector, b, one computes A*b, then one multiplies that vector by A to find A^2b and so on. Because the vectors usually soon become almost linearly dependent due to the properties of power iteration, methods relying on Krylov subspace frequently involve some orthogonalization scheme such as Arnoldi's iteration for more general matrices.
- **Arnoldi's iteration**: It is an eigenvalue algorithm. Also, an example of an iterative method and is an analogy for computing the QR factorization A = QR of a matrix A. Arnoldi's iteration finds an approximation to the eigenvalues and eigenvectors of general matrices by forming an orthonormal basis of the Krylov subspace via Gram-Schmidt Orthogonalization, which is useful when dealing with large sparse matrices.

2.5.3 Hybrid Methods for Solving SLEs

Hybrid methods combine direct and iterative methods to exploit the advantages of both direct and iterative methods. It depends on partitioning the matrix and solving its subsets using direct or iterative methods [75]. In [76], the authors proposed a hybrid approach, where the matrix is partitioned into blocks. Within each block, a highly optimized (parallel) conventional solver is used, then the blocks are coupled together using block Jacobi. This allows limiting the block size to the point where the conventional iterative methods no longer scale. Moreover, several combinations of evolutionary computation techniques and classical numerical methods are proposed to solve linear equations.

2.5.4   ML-Based Methods for Solving SLEs

It enhances solving SLEs by using artificial intelligence techniques to accelerate the convergence of iterative numerical methods. The proposed method will be based on genetic algorithm (GA) and neural network (NN) artificial intelligence methods [47, 75–88]. These methods have the potential to significantly accelerate large-scale computational efforts by accelerating the convergence of iterative numerical methods. GA and NN SLEs solver were able to find all possible sets of solutions that are applicable to any given system of linear equations. Conventional methods always produce a set of solutions for a particular system of linear equations. Noe that the population random values should be in the A and b range.

2.5.5   Quantum-Based Methods for Solving SLEs

Variational algorithms offer a promising avenue for tackling linear algebra problems efficiently. These algorithms leverage variational principles to approximate solutions to linear systems, matrix inversion, and other algebraic tasks. By formulating the problem as an optimization task, variational methods seek to minimize a suitable objective function, often representing the discrepancy between the true solution and an approximate solution. One prominent example is variational quantum algorithms for linear algebra, which have garnered significant attention in recent years. These algorithms harness the power of quantum computing to perform linear algebra computations by encoding the problem into a quantum state and utilizing quantum operations to manipulate the state towards the desired solution. Variational quantum algorithms hold promise for solving linear systems, computing matrix inverses, and other linear algebra tasks with potentially improved efficiency compared to classical counterparts [89]. This algorithm is a promising solution for accelerating computations and solving large-scale problems efficiently.

2.5.6   How to Choose a Method for Solving Linear Equations

As stated earlier, there exists a variety of algorithms for solving linear systems (Fig. 2.16). Selecting the best solver algorithm depends on the matrix structure as well as on different trade-offs such as computational complexity, memory bottlenecks, convergence properties, and numerical behavior. Comparison between different methods based on the symmetric 3×3 equations in terms of computation time and number of iterations is shown in Table 2.10.

Direct solvers suffer from slow convergence for large systems of equations. Moreover, it takes more time to converge for sparse linear system as compared to dense linear system. So, iterative methods are preferred in that case.

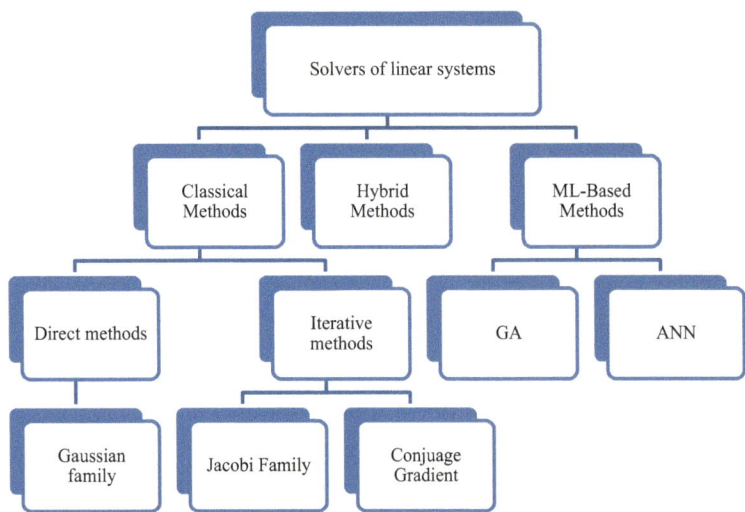

Fig. 2.16   Different methods of solving systems of linear equations

Table 2.10   Comparison between different iterative methods

Methods	Computation time (s)	Number of iterations	Memory usage
Jacobi	0.8	10	The highest
Gauss–Seidel	0.75	9	High
SOR	0.7	8	Average
Conjugate gradient	0.05	6	The lowest
Gaussian elimination	0.03	–	Low
Gauss-Jordan	0.03	–	Low
LU factorization	0.02	–	Low

The simulation results show that the SOR method converges faster than the Jacobi and Gauss–Seidel methods. Jacobi method is the slowest one in terms of speed and the highest one in terms of usage of memory. Conjugate gradient is the fastest among all iterative and direct solvers. But, it has one restriction that the matrix should be symmetric. For direct solver, LU Factorization is the best. The solutions are very accurate for all methods.

The criteria on which we can choose which method to be used are summarized in Table 2.11. The two important criteria to take into consideration when choosing a method for solving linear equations are: convergence rate and the cost of calculating the method.

Table 2.11 How to choose a method for solving linear equations

SLEs solver methods		Matrix type					
		Dense				Sparse	
		Square		Not square		Symmetric	Not symmetric
		Symmetric	Not symmetric				
Direct solver	Cramer	✓	✓	✗		✗	
	Gaussian elimination	✓	✓	✓			
	Gauss-Jordan elimination	✓	✓	✓			
	LU decomposition	✓	✓;	✓			
	Cholesky decomposition	✓	✗	✗			
Iterative/ indirect solver	Stationary methods	Gauss–Seidel	✗				✓	✓
		Jacobi					✓	✓
		Gaussian over relaxation					✓	✓
	Krylov methods	Conjugate gradient					✓	✗
		Bi-conjugate gradient					✓	✗
		GMRES					✗	✓

2.6 Different Methods for MOR

The model order reduction (MOR) techniques are falling under four main categories as illustrated in Fig. 2.17. The comparison between these different MOR methods is shown in Table 2.12. MOR represents an important research area which has been widely applied not only for electromagnetic problems, but also in other domains such as fluid dynamics, mechanics, computational biology, circuit design, control theory, biomedical applications, and so on [90]. Despite all existing MOR techniques, many unsolved problems still remain. Moreover, nonlinear MOR (NMOR) [91], and parameterized MOR (PMOR) [92], still not mature.

For NMOR, transferring these methods to the case of nonlinear systems is not straight-forward. For PMOR, the high-order systems depend on parameters, for example geometry or material parameters. Reducing the order of a large-scale system and at the same time preserving the parametric dependencies is a challenge. Moreover, there is no method that gives the best results for all of the systems. So, each system uses the best method according to its application. So, there is still a need for novel MOR techniques.

Moreover, machine learning based algorithms such as: genetic algorithm, artificial neural networks, Fuzzy logic, Particle swarm optimization, and simulated annealing have

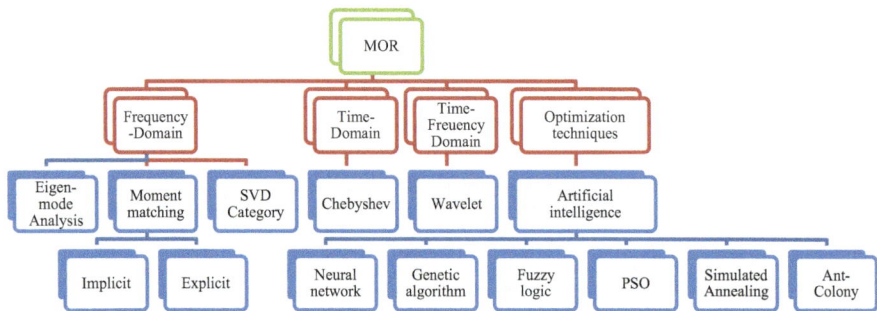

Fig. 2.17 MOR categories

shown some promising results in solving model order reduction problem [93]. MOR is a critical enabler for data-driven learning.

2.7   Common Hardware Architecture for Different Numerical Solver Methods

Numerical solver methods can be implemented on a variety of hardware architectures, depending on the specific requirements of the problem and the available resources. However, there are some common hardware architectures that can be used to implement different numerical solver methods, including:

- Central Processing Unit (CPU): A CPU is a common hardware architecture used in numerical solvers. It is a general-purpose processor that can be programmed to perform a wide variety of tasks, including numerical computations. CPUs are typically designed to support single-threaded or multi-threaded processing, with each thread executing a different part of the code simultaneously.
- Graphics Processing Unit (GPU): GPUs are specialized hardware architectures designed to accelerate the processing of graphics and other highly parallelizable tasks, including numerical computations. GPUs can be used to accelerate numerical solvers that require a large number of computations to be performed in parallel, such as those used in machine learning and scientific computing.
- Field-Programmable Gate Array (FPGA): FPGAs are programmable logic devices that can be used to implement custom hardware architectures tailored to specific numerical solver methods. They can be used to accelerate the processing of numerical computations by offloading them from the CPU or GPU and can be reprogrammed to adapt to changing requirements or to optimize performance for a specific problem.

Table 2.12 The comparison between different MOR methods

Method	How it works	Advantage	Disadvantage
Frequency-domain			
Eigen-mode analysis	For linear systems only Pole-residue form: $$H(s) = \sum_{i=1}^{n} \frac{R_i}{s - p_i}$$ Pole-zero form: $$H(s) = \frac{\prod_{i=1}^{n-1} (s - \zeta_i)}{\prod_{i=1}^{n} (s - p_i)}$$ Rational form (point-matching): $$H(s) =$$ $$\frac{b_0 + b_1 s + \cdots + b_{N-1} s^{N-1}}{1 + a_1 s + \cdots + a_N s^N}$$ b_i represents the numerator coefficients, a_i represents the denominator coefficients	• Drop terms with small residues • Drop terms with large poles • Find a low order rational function matching • Take the part of the transfer function with the poles that are the closest to the imaginary axis and to throw away the others	Simple physical interpretation	Computationally inefficient

(continued)

Table 2.12 (continued)

Method	How it works	Advantage	Disadvantage
Moment matching category	AWE (implicit) $$H(s) = \sum_{i=0}^{\infty} m_i(s - s_0)^i$$ With respect to $s_0 = 0$ $$= m_0 + m_1 s + m_2 s^2 + \cdots$$	Taylor expansion	Simple and computationally efficient well-conditioned	Low accuracy Does not guarantee passivity or stability
	Padé approximations $$\frac{\hat{b}_0+\hat{b}_1 s+\cdots+\hat{b}_{q-1}s^{q-1}}{1+\hat{a}_1 s+\cdots+\hat{a}_q s^q} = m_0 + m_1 s + m_2 s^2 + \cdots + m_{2q-1}s^{2q-1}$$	Rational function fitting via moment matching	Improve the accuracy for low frequency Can handle **MIMO** problems	Ill-conditioning. (Narrow band), leading to inaccuracies at higher orders ODEs
	Krylov subspace-scheme (explicit): (Arnoldi, PVL, PRIMA)	Projection (change of variables)	Improve the ill-conditioning (wide-band)	Cannot handle MIMO problems Stability not preserved No Global error bound
SVD category	Proper orthogonal decomposition (POD) method (nonlinear system)	Decomposition	Improve the ill-conditioning	Computationally inefficient So it is used with low-order systems
	Balanced truncation (linear)	Truncation	Improve the ill-conditioning	Computationally inefficient So it is used with low-order systems

(continued)

Table 2.12 (continued)

Method		How it works	Advantage	Disadvantage
	Hankel approximation (linear)	Approximation	Improve the ill-conditioning	Computationally inefficient So it is used with low-order systems
Time-domain				
Chebyshev		In time domain	Preserve passivity and stability	Error increases with high-order of the Chebyshev polynomial
Time-frequency domain				
Wavelet		In time frequency domain	Computationally efficient	–
Optimization techniques				
Machine learning	Neural network	This will be discussed in this book		
	Genetic algorithm			
	Fuzzy logic			
	Particle swarm optimization (PSO)			
	Simulating annealing			
	Ant-colony			

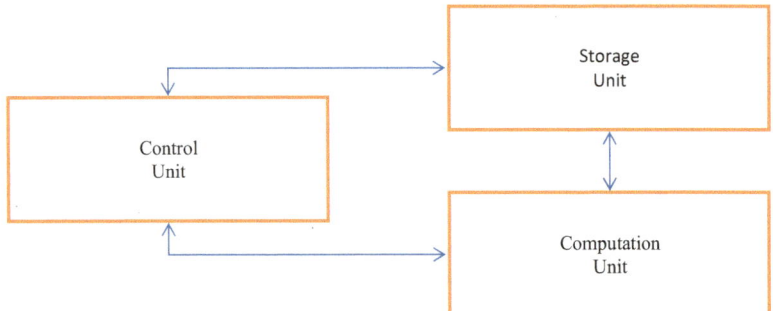

Fig. 2.18 Block diagram describing the general structure of the proposed generic Hardware-BASED numerical analysis solver

- Application-Specific Integrated Circuit (ASIC): ASICs are specialized hardware architectures designed to perform specific tasks efficiently. They can be used to implement numerical solver methods that require high-performance and low-power consumption, such as those used in embedded systems or mobile devices.

Overall, the choice of hardware architecture depends on a variety of factors, including the specific numerical solver method being implemented, the size and complexity of the problem, and the available resources. CPU and GPU architectures are limited by their maximum memory and computational bandwidth which are considered low compared to FPGA for these types of problems. In this section, generalized FPGA-based hardware architecture for different methods used to solve system of linear equations is presented to speed-up the solving time. We build a complete hardware library of most of the famous methods with many shared hardware blocks. The proposed architecture consists of a control unit, storage unit, and computation unit. Block diagram describing the general structure of the proposed generic hardware-Based numerical analysis solver is shown in Fig. 2.18. The proposed methodology is summarized in Fig. 2.19. The proposed parallelized architecture can speed-up the execution time for many applications over software solutions.

2.8   Software Implementation for Different Numerical Solver Methods

Numerical solver methods can be implemented in a variety of software environments, depending on the specific requirements of the problem and the available resources. Here are some common software implementation options [94–103]:

Fig. 2.19 The proposed
methodology

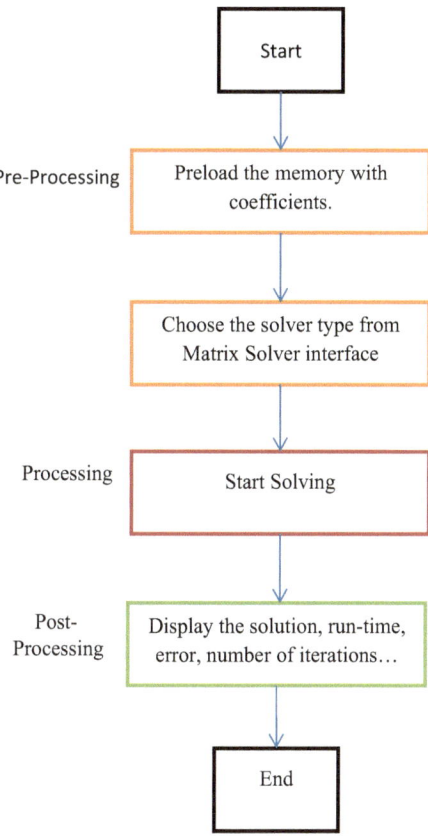

- General-purpose programming languages: Numerical solver methods can be implemented in general-purpose programming languages such as Python, C++, Java, and Fortran. These languages provide a wide range of libraries and tools for numerical computation and scientific computing, making them a popular choice for implementing numerical solver methods.
- Numerical computing libraries: There are several numerical computing libraries available that provide efficient implementations of numerical solver methods. Examples of these libraries include NumPy, SciPy, MATLAB, and Mathematica. These libraries can significantly simplify the implementation of numerical solver methods and provide a high-level interface for solving complex problems.
- Domain-specific languages: Domain-specific languages (DSLs) are programming languages designed specifically for a particular domain, such as numerical computation. DSLs can be used to implement numerical solver methods and provide a high-level interface that is tailored to the specific problem domain.

- Integrated development environments (IDEs): IDEs such as MATLAB, Octave, and Spyder provide a complete development environment for implementing and testing numerical solver methods. These environments provide tools for debugging, profiling, and visualizing the results of the solver methods.

Overall, the choice of software implementation depends on a variety of factors, including the specific numerical solver method being implemented, the size and complexity of the problem, and the available resources.

2.8.1 Cramer's Rule: Python-Implementation

Cramer's Rule

```python
import numpy as np

def det2x2(A, v=False):
    if v:  print 'compute 2 x 2 det of'
    if v:  print A
    assert A.shape == (2,2)
    return A[0][0]*A[1][1] - A[0][1]*A[1][0]

def det3x3(A):
    print 'compute 3 x 3 det of'
    print A
    assert A.shape == (3,3)
    a,b,c = A[0]
    c1 = a * det2x2(A[1:3,[1,2]])
    c2 = b * det2x2(A[1:3,[0,2]])
    c3 = c * det2x2(A[1:3,[0,1]])
    return c1 - c2 + c3

def solve(A):
    print 'solve'
    print A, '\n'
    assert A.shape == (3,4)
    D = det3x3(A[:,:3])
    print 'D = ', D, '\n'
    if D == 0:
        print 'no solution'
        return
    Dx = det3x3(A[:,[3,1,2]])
    print 'Dx = ', Dx, '\n'
    Dy = det3x3(A[:,[0,3,2]])
    print 'Dy = ', Dy, '\n'
    Dz = det3x3(A[:,[0,1,3]])
    print 'Dz = ', Dz, '\n'
    return Dx*1.0/D, Dy*1.0/D, Dz*1.0/D
```

2.8.2   Newton Raphson: C-Implementation

Newton Raphson

```
// C++ program for implementation of Newton Raphson Method for
// solving equations
#include<bits/stdc++.h>
#define EPSILON 0.001
using namespace std;
// An example function whose solution is determined using
// Bisection Method. The function is x^3 - x^2  + 2
double func(double x)
{
   return x*x*x - x*x + 2;
}

// Derivative of the above function which is 3*x^x - 2*x
double derivFunc(double x)
{
   return 3*x*x - 2*x;
}

// Function to find the root
void newtonRaphson(double x)
{
   double h = func(x) / derivFunc(x);
   while (abs(h) >= EPSILON)
   {
      h = func(x)/derivFunc(x);

      // x(i+1) = x(i) - f(x) / f'(x)
      x = x - h;
   }

   cout << "The value of the root is : " << x;
}

// Driver program to test above
int main()
{
   double x0 = -20; // Initial values assumed
   newtonRaphson(x0);
   return 0;
}
```

2.8.3   Gauss Elimination: Python-Implementation

Gauss Elimination

```
def myGauss(m):
    #eliminate columns
    for col in range(len(m[0])):
        for row in range(col+1, len(m)):
            r = [(rowValue * (-(m[row][col] / m[col][col]))) for rowValue in m[col]]
            m[row] = [sum(pair) for pair in zip(m[row], r)]
    #now backsolve by substitution
    ans = []
    m.reverse() #makes it easier to backsolve
    for sol in range(len(m)):
        if sol == 0:
            ans.append(m[sol][-1] / m[sol][-2])
        else:
            inner = 0
            #substitute in all known coefficients
            for x in range(sol):
                inner += (ans[x]*m[sol][-2-x])
            #the equation is now reduced to ax + b = c form
            #solve with (c - b) / a
            ans.append((m[sol][-1]-inner)/m[sol][-sol-2])
    ans.reverse()
    return ans
```

2.8.4   Conjugate Gradient: MATLAB-Implementation

Conjugate Gradient

```
close all;
clear all;
clc;
prompt = 'Please Enter the A Matrix, taking into consideration A being symmetric & positive: ';
A = input(prompt)
prompt = 'Pleae Enter The b Vector';
b = input(prompt)
prompt = 'Please Enter Initial guess: ';
x = input(prompt);
r = b-A*x;
p = r;
rcomp = r'*r
 for i = 1:length(b)
     Ap = A * p;
     alpha = rsold / (p' * Ap);
     x = x + alpha * p;
     r = r - alpha * Ap;
     rsnew = r' * r;
     if sqrt(rsnew) < 1e-10
         break;
     end
     p = r + (rsnew / rsold) * p;
     rsold = rsnew;
 end
```

2.8.5   GMRES: MATLAB-Implementation

GMRES

```
%% Initialization
clc
clear all
close all

%% Code
A = [3 4 5; 6 7 8; 9 10 11]; %nxn matrix
b = [1 ;1 ;1];          %nx1 vector
n=size(A,1);             %To get its size
x0 = [-4 ;5 ;-9];        %Arbitrary guess
r0 = b - A*x0;           %Our guess
k = 5;                   %Number of iterations,
                         %If we increase k we increase
                         %accuracy but computational time increases.

V=zeros(n,k+1);          %Our n x k+1 matrix
H=zeros(k+1,k);          %Our upper matrix.
V(:,1)=r0/norm(r0);      %Our initial guess of V.

% Arnoldi, The Gram-Schmidt implementation
for j=1:k
    z=A*V(:,j);
    for i=1:j
        H(i,j)=dot(z,V(:,i)); %Dot product
        z=z-H(i,j)*V(:,i);
    end
    H(j+1,j)=norm(z);
    if H(j+1,j)==0, break, end
    V(:,j+1)=z/H(j+1,j);
end
```

2.8.6   Cholesky: MATLAB-Implementation

Cholesky

```
%% Initialization
clc
clear all
close all

%% Code
A = [3 4 5; 6 7 8; 9 10 11]; %nxn matrix
b = [1 ;1 ;1];          %nx1 vector
n=size(A,1);            %To get its size
x0 = [-4 ;5 ;-9];       %Arbitrary guess
r0 = b - A*x0;          %Our guess
k = 5;                  %Number of iterations,
                        %If we increase k we   increase
                        %accuracy but computational time increases.

V=zeros(n,k+1);         %Our n x k+1 matrix
H=zeros(k+1,k);         %Our upper matrix.
V(:,1)=r0/norm(r0);     %Our initial guess of V.

% Arnoldi, The Gram-Schmidt implementation
for j=1:k
    z=A*V(:,j);
    for i=1:j
        H(i,j)=dot(z,V(:,i)); %Dot product
        z=z-H(i,j)*V(:,i);
    end
    H(j+1,j)=norm(z);
    if H(j+1,j)==0, break, end
    V(:,j+1)=z/H(j+1,j);
end
```

2.9   Stochastic Differential Equations

Stochastic Differential Equations (SDEs) are a class of differential equations in which one or more terms are driven by stochastic processes, typically modeled with Brownian motion or Wiener processes. They are used to describe systems or processes that exhibit randomness or uncertainty, often found in fields such as finance, physics, biology, and engineering. SDEs often do not have closed-form analytical solutions, so numerical methods are used to approximate solutions. Popular methods include [104, 105]:

- **Euler–Maruyama Method**: A simple extension of the Euler method for ODEs, suitable for basic SDE simulations.
- **Milstein's Method**: An improvement on Euler–Maruyama that accounts for higher-order terms, providing more accuracy.

- **Runge–Kutta Methods**: Adaptations of the deterministic Runge–Kutta methods to SDEs.

Their applications can include:

- **Finance**: SDEs model asset prices, interest rates, and other financial instruments (e.g., the Black–Scholes model for option pricing).
- **Physics**: Used to describe systems with random fluctuations, such as particles in a fluid.
- **Biology**: Modeling population dynamics, gene expression, or neural activity with inherent randomness.
- **Engineering**: Describing systems with noise, such as in control theory or signal processing.

2.10   Conclusions

In this chapter, we introduce the numerical analysis methods for electronics. Numerical analysis is the study of algorithms that use numerical approximation for the problems of mathematical analysis. Numerical analysis naturally finds application in all fields of engineering and the physical sciences. Also, the life sciences, social sciences, medicine, business and even the arts have adopted elements of scientific computations.

In this chapter, we introduce different approaches to solve partial differential equations (PDEs) and ordinary differential equations (ODEs) as well as the advantages and disadvantages of each method are analyzed. In this chapter, we introduce different approaches to solve system of nonlinear equations (SNLEs) as well as the advantages and disadvantages of each method are analyzed. In this chapter, we introduce different approaches to solve system of linear equations (SLEs) as well as the advantages and disadvantages of each method are analyzed.

References

1. De Luca, G., Antonini, G., Benner, P. (2013). A parallel, adaptive multi-point model order reduction algorithm. In *22nd IEEE Conference on Electrical Performance of Electronic Packaging and Systems (EPEPS)*.
2. Wang, C., Yu, H., Li, P., Ding, C., Sun, C., Guo, X., Zhang, F., Zhou, Y., & Yu, Z. (2013) Krylov subspace based model reduction method for transient simulation of active distribution grid. In *IEEE Power and Energy Society General, Meeting (PES), PESMG*.
3. Gould, H., Tobochnik, J., & Christian, W. (2007). *An introduction to computer simulation methods: Applications to physical systems* (3rd ed.). Pearson/Addison-Wesley.
4. Gould, H., & Tobochnik, J. (1996). *An introduction to computer simulation methods* (2nd ed.). Addison-Wesley.

5. Wong, S. S. M. (1997). *Computational methods in physics and engineering.* World-Scientific.
6. Press, W. H., et al. (1992). *Numerical recipes in FORTRAN: The art of scientific computing.* Cambridge University Press.
7. Press, W. H., et al. (1988). *Numerical recipes in C: The art of scientific computing.* Cambridge University Press.
8. Pang, T. (1997). *An introduction to computational physics.* Cambridge University Press for program listings in Fortran 90 and Mathematica.
9. Garcia, A. L. (2000). *Numerical methods for physics* (2nd ed.). Prentice Hall for program listings in MatLab and C++.
10. Giordano, N. J. (1997). *Computational physics.* Prentice Hall for program listings in True Basic.
11. Salah, K. (2017). A novel model order reduction technique based on artificial intelligence. *Microelectronics Journal, 65,* 58–71.
12. Gupta, S. K. (1995). *Numerical Methods for Engineers.* Wiley Eastern.
13. Hager, W. W. (1988). *Applied numerical linear algebra.* Prentice Hall.
14. Dongarra, J. J., Du, I. S., Sorensen, D. C., & van der Vorst, H. A. (1998). *Numerical linear algebra for high-performance computers.* SIAM.
15. Hoffman, J. D. (1992). *Numerical methods for engineers and scientists.* McGraw-Hill Inc.
16. Raghavan, P., Teranishi, K., & Ng, E. G. (2003). A latency tolerant hybrid sparse solver using incomplete Cholesky factorization. *Numerical Linear Algebra and Applications, 10,* 541–560.
17. Mohamed, K. S. (2020). Numerical computing. In *Neuromorphic computing and beyond.* Springer. https://doi.org/10.1007/978-3-030-37224-8_2
18. Gerez, S. H. (1999). *Algorithms for VLSI design automation.* Wiley.
19. Wang, L.-T., Chang, Y.-W., & Cheng, K.-T. T. (2009). Electronic design automation: Synthesis, verification, and test. *The Morgan Kaufmann series in systems on silicon.* Morgan Kaufmann.
20. Jansen, D. (2003). *The electronic design automation handbook.* Kluwer Academic Publishers.
21. Alpert, C. J., Mehta, D. P., & Sapatnekar, S. S. (2009). *Handbook of algorithms for physical design automation.* CRC Press.
22. Kahng, A. B., Lienig, J., Markov, I. L., & Hu, J. (2011). *VLSI physical design: From graph partitioning to timing closure.* Springer.
23. Sait, S. M., & Youssef, H. (1999). *VLSI physical design automation: Theory and practice.* Lecture notes series on computing (Vol. 6). World Scientific Publishing.
24. Sarrafzadeh, M., & Wong, C. K. (1996). *An introduction to VLSI physical design.* McGraw-Hill series in computer science. McGraw-Hill.
25. Sherwani, N. A. (1999). *Algorithms for VLSI physical design automation* (3rd ed.). Kluwer Academic Publishers.
26. Chang, C. H., Chen, Y. J., & Huang, W. C. (2019). An overview of electronic design automation. *IEEE Access, 7,* 35065–35078.
27. Pinto, P. S., & Silva, J. A. C. (2019). Electronic design automation: a survey of tools and techniques. *IEEE Design & Test, 36*(1), 49–61.
28. Zhang, R., & Stan, M. R. (2019). The role of electronic design automation in emerging technologies. *Proceedings of the IEEE, 107*(1), 57–73.
29. Venkatesh, S., & Mohapatra, S. S. (2019). Electronic design automation for internet of things (IoT) applications. *IEEE Transactions on Computer-Aided Design of Integrated Circuits and Systems, 38*(9), 1561–1574.
30. Sarrafzadeh, A., & Pedram, M. (2019). Electronic design automation for low-power systems-on-chip. *Proceedings of the IEEE, 107*(4), 716–727.

31. https://www.scratchapixel.com/lessons/mathematics-physics-for-computer-graphics/monte-carlo-methods-in-practice/monte-carlo-methods.html
32. James, G. (2007). *Modern engineering mathematics* (4th ed., p. 778). Prentice Hall.
33. James, G. (2011). *Advanced modern engineering mathematics* (4th ed., p. 164). Prentice Hall.
34. https://en.wikipedia.org/wiki/Finite_difference_method
35. Strang, G., & Fix, G. (1973). *An analysis of the finite element method*. Prentice Hall.
36. Polycarpou, A. C. (2006). *Introduction to the finite element method in electromagnetics*. Morgan & Claypool.
37. Farmani, A. (2019). Three-dimensional FDTD analysis of a nanostructured plasmonic sensor in the near-infrared range. *Journal of the Optical Society of America B, 36*(2), 401.
38. Zeng, Z., Venuthurumilli, P. K., Xu, X. (2021). Inverse design of plasmonic structures with FDTD. *ACS Photonics, 8*(5), 1489–1496.
39. Ruehli, A. (1996). Partial element equivalent circuit (PEEC) method and its application in the frequency and time domain. In *Proceedings of Symposium on Electromagnetic Compatibility* (pp. 128–133). https://doi.org/10.1109/ISEMC.1996.561214
40. http://mathworld.wolfram.com/LegendrePolynomial.html
41. http://mathworld.wolfram.com/Fourier-LegendreSeries.html
42. Cole, K. D., Beck, J. V., Haji-Sheikh, A., & Litkouhi, B. (2011). *Heat conduction using Green's functions* (2nd ed.). CRC Taylor and Francis.
43. Rabiei, F. A., Ismail, F., & Suleiman, M. (2013). Improved Runge-Kutta methods for solving ordinary differential equations. *Sains Malaysiana, 42*(11), 1679–1687.
44. Diaz-Toca, G. M., GonzalezVega, L., & Lombardi, H. (2005). Generalizing Cramers rule: Solving uniformly linear systems of equations. *SIAM Journal of Matrix Analysis and Applications, 27*, 621–637.
45. Hussian, A. (2015). Numerical solution of partial differential equations by using modified artificial neural network. *Network and Complex Systems, 5*(6).
46. He, J., Xu, J., & Yao, X. (2000). Solving equations by hybrid evolutionary computation techniques. *IEEE Transactions on Evolutionary Computation, 4*(3).
47. Shahzadehfazeli, S. A. (2016). An improved iterative method for solving general system of equations via genetic algorithms. *International Journal of Information Technology, Modeling and Computing (IJITMC), 4*(1).
48. Barry-Straume, J., et al. (2022). Physics-informed neural networks for PDE-constrained optimization and control. arXiv preprint arXiv:2205.03377
49. Antonelo, E. A., Camponogara, E., Seman, L. O., de Souza, E. R., Jordanou, J. P., & Hubner, J. F. (2021). Physics-informed neural nets for control of dynamical systems.
50. Zhang, D., Guo, L., & Karniadakis, G. E. (2019). *Learning in modal space: Solving time-dependent stochastic PDEs using physics-informed neural networks.*
51. http://www.ugrad.math.ubc.ca/coursedoc/math100/notes/approx/newton.html
52. https://www.ijser.org/researchpaper/Newton-Raphson-Method.pdf
53. http://www.seas.ucla.edu/~vandenbe/236C/lectures/qnewton.pdf
54. https://www.stat.cmu.edu/~ryantibs/convexopt/lectures/quasi-newton.pdf
55. The official University of British Columbia Math Division website: https://www.math.ubc.ca/~pwalls/math-python/roots-optimization/secant/
56. The open-source java codes website: http://theflyingkeyboard.net/java/java-secant-method-2/
57. https://www.geeksforgeeks.org/program-muller-method/
58. Flyer, N. (2008). A hybrid analytical–numerical method for solving evolution partial differential equations. I. The half-line. *Proceeding of the Royal Society.*
59. Dhamacharon, A. (2014). An efficient hybrid method for solving systems of nonlinear equations. *Journal of Computational and Applied Mathematics, 263*, 59–68.

60. Luo, Y. Z. (2008). Hybrid approach for solving systems of nonlinear equations using chaos optimization and quasi-Newton method. *Applied Soft Computing, 8*(2).
61. Fathalizadeh, H. (2016). Solving nonlinear ordinary differential equations using neural networks. In *2016 4th International Conference on Control, Instrumentation, and Automation (ICCIA)*, January 27–28, 2016
62. Joshi, G. (2014). Solving system of non-linear equations using genetic algorithm. In *International Conference on Advances in Computing, Communications and Informatics (ICACCI)*.
63. Habgood, K., & Arel, I. (2012). A condensation-based application of Cramers rule for solving large-scale linear systems. *Journal of Discrete Algorithms*.
64. Hochet, B., Quinton, P., & Robert, Y. (1989). Systolic Gaussian elimination overGF(p)with partial pivoting. *IEEE Transactions on Computers, 38*(9), 1321–1324.
65. http://caslab.csl.yale.edu/code/gausselim/
66. Kyrchei, I. (2012). Cramer's rule for quaternionic systems of linear equations. *Journal of Mathematical Sciences, 155*, 839–858.
67. https://www.austincc.edu/jthom/GaussJordanElimination.pdf
68. https://www.codewithc.com/c-program-for-gauss-jordan-method/
69. Press, W. H., Teukolsky, S. A., Vetterling, W. T., & Flannery, B. P. (2007). *Numerical recipes (The art of scientific computing)* (3rd ed.). Cambridge University Press.
70. Burden, R. L., & Faires, J. D. (2005). *Numerical analysis* (8th ed.). Thomson Learning Inc.
71. Moussa, S., Razik, A. M. A., Dahmane, A. O., & Hamam, H. (2013). FPGA implementation of floating-point complex matrix inversion based on GAUSS-JORDAN elimination. In *26th Annual IEEE Canadian Conference on Electrical and Computer Engineering (CCECE)*, Regina, Canada, May 2013.
72. Duarte, R., Neto, H., & Vestias, M. (2009). Double-precision Gauss-Jordan algorithm with partial pivoting on FPGAs. In *12th Euromicro Conference on Digital System Design, Architectures, Methods and Tools (DSD)*, August 2009 (pp. 273–280).
73. https://www.maa.org/press/periodicals/loci/joma/iterative-methods-for-solving-iaxi-ibi-jac obis-method
74. Yang, D., et al. (2010). Performance comparison of Cholesky decomposition on GPUs and FPGAs. In *Symposium on Application Accelerators in High Performance Computing*.
75. https://en.wikipedia.org/wiki/Krylov_subspace
76. https://en.wikipedia.org/wiki/Arnoldi_iteration
77. Morris, G. R., & Abed, K. H. (2013). Mapping a Jacobi iterative solver onto a high performance heterogeneous computer. *IEEE Transactions on Parallel and Distributed Systems, 24*(1), 85–91.
78. Brown, N. (2013). *Solving large sparse linear systems using asynchronous multi-splitting*. PRACE.
79. Gerald, C. F., & Wheatley, P. O. (1998). *Applied numerical analysis* (5th ed.). Addison-Wesley.
80. Kelley, C. (1995). *Iterative methods for linear and nonlinear equations*. Society for Industrial Mathematics.
81. Salah, K. (2017). A generic model order reduction technique based on Particle Swarm Optimization (PSO) algorithm. In *IEEE EUROCON*.
82. http://mathforcollege.com/nm/mws/gen/04sle/mws_gen_sle_txt_seidel.pdf
83. http://www.preyeshdalmia.com/assets/Iterative_solver.pdf
84. https://www.sanfoundry.com/cpp-program-implement-gauss-seidel-method/
85. https://github.com/nuhferjc/gauss-seidel/blob/master/gaussseidel.py
86. https://matrix.reshish.com/cramer.php
87. https://perso.uclouvain.be/paul.vandooren/Krylov.pdf

88. Hochet, B., Quinton, P., & Robert, Y. (1989). Systolic Gaussian elimination over GF(p) with partial pivoting. *IEEE Transactions on Computers, 38*(9), 1321–1324.

89. Xu, X., Sun, J., Endo, S., Li, Y., Benjamin, S. C., & Yuan, X. (2021). Variational algorithms for linear algebra. *Science Bulletin, 66*, 2181–2188.

90. Ramirez, A. (2014). Reduced-order state-space systems in the dynamic harmonic domain. In *16th IEEE International Conference on Harmonics and Quality of Power (ICHQP)*.

91. Aridhi, H., Zaki, M. H., & Tahar, S. (2012). Towards improving simulation of analog circuits using model order reduction. In *Design, Automation & Test in Europe Conference & Exhibition (DATE)*.

92. Geuss, M., Panzer, H., & Lohmann, B. (2013). On parametric model order reduction by matrix interpolation. In *European Control Conference (ECC)*, Zürich, Switzerland, July 17–19, 2013.

93. Mohamed, K. S. (2018). *Machine learning for model order reduction* (Vol. 664). Springer.

94. Higham, N. J. (2002). *Accuracy and stability of numerical algorithms*. Society for Industrial and Applied Mathematics.

95. Heath, M. T. (2002). *Scientific computing: an introductory survey*. McGraw-Hill Education.

96. Press, W. H., Teukolsky, S. A., Vetterling, W. T., & Flannery, B. P. (2007). *Numerical recipes: The art of scientific computing*. Cambridge University Press.

97. Lippert, T., et al. (2008). Towards exascale computing for lattice QCD. *Journal of Physics: Conference Series, 125*(1), 012025.

98. Dongarra, J., et al. (2009). The international exascale software project roadmap. *International Journal of High Performance Computing Applications, 23*(3), 309–322.

99. McKinley, R. (2012). The role of high-level languages in numerical computation. *ACM SIGPLAN Notices, 47*(7), 51–56.

100. Van Rossum, G., & Drake, Jr., F. L. (2011). *Python 3 reference manual*. CreateSpace.

101. Oliphant, T. E. (2007). Python for scientific computing. *Computing in Science & Engineering, 9*(3), 10–20.

102. Shampine, L. F., & Thompson, S. (2001). Solving DAEs in MATLAB. *Applied Numerical Mathematics, 37*(4), 441–458.

103. Kubatko, E. J., & Kim, K. H. (2006). *Numerical methods for chemical engineering: Applications in MATLAB*. CRC Press.

104. Simo Särkkä and Arno Solin. (2019). *Applied stochastic differential equations*. Cambridge University Press.

105. Hairer, E., Nørsett, S., & Wanner, G. (1993). Solving ordinary differential equations I: Nonstiff problems. Springer. ISBN: 978-3-540-56670-0.

The Basics of EDA Tools for IC

"A Physics-Aware Approach"

3.1   Introduction: System Classification

A **system** is a group of components that have specific attributes that are combined to form certain relations to perform one or few specific functions. This definition is constructed upon four concepts:

- **Components**: the elementary material parts or the building blocks of the system. According to the role in the system, the different components can be classified into one of the following categories:
  - **Structural components**: The static or quasi-static parts of the system.
  - **Operating components**: The parts that perform the processing (dynamic).
  - **Flow components**: material, data or energy processed by the systems.
- **Attributes**: the properties of the components.
- **Relations**: the connections between components.
- **Functions**: what the system needs to accomplish.

Systems can be classified according to their own characteristics. General classes of systems include (Table 3.1):

- Natural and human-made.
- Static and dynamic: Dynamic systems can have several subclasses:
  - linear or nonlinear.
  - discrete time or continuous time.
  - Periodic or event-driven.
  - deterministic or stochastic (adaptive).
  - single input or multiple input.

© The Author(s), under exclusive license to Springer Nature Switzerland AG 2025   91
K. S. Mohamed, *Next Generation EDA Flow*, Synthesis Lectures on Engineering,
Science, and Technology, https://doi.org/10.1007/978-3-031-88435-1_3

Table 3.1 Important properties of systems

Property	Description
Isotropic (scalar)	Direction independent
Homogenous	Position independent $\frac{d}{dx} = 0$
Time-invariant	Its response to the input does not change with time
Static (steady-state)	Time independent $\frac{d}{dt} = 0$
Dynamic	Their behavior depends on their past evolution
Linear	Field independent $! = f(v)$ Follows the principle of superposition
Stationary	Frequency does not change with time
Causal	Future-input independent, depends only on current and past values
Deterministic	Not stochastic
Ergodic	The output is not repeated in a certain range
Ubiquitous	Provides context-aware services ubiquitously (Anywhere/anytime)

- single output or multiple output.
- stable or unstable (chaotic).
- Closed and open: Systems that have negligible interactions with their environments are called closed systems.

Simulations can be deterministic or stochastic (Monte-Carlo).

3.1.1   Problem Solving

Problems themselves can be classified into two different categories known as ill-defined and well-defined problems (Fig. 3.1). Before finding a solution to the problem, the problem must first be clearly identified. There are many strategies for problem solving (Table 3.2) [1]. An algorithm is a set of instructions to do a specified task. Algorithms are vital especially for processing big data. Problem solving steps [2]:

- Identify a problem.
- Understand the problem.
- Identify alternative ways to solve a problem.
- Select the best way to solve a problem from the list of alternative solutions.
- Evaluate the solution.

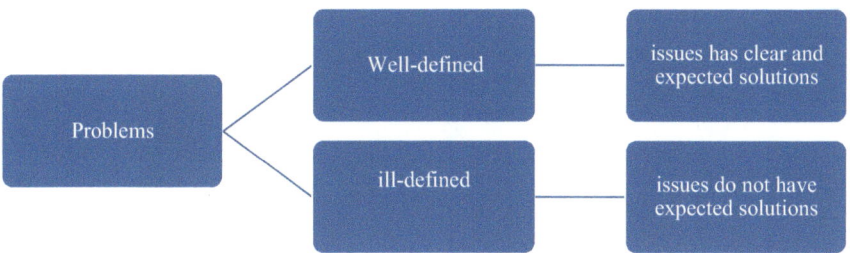

Fig. 3.1 Problem classifications

Table 3.2 Problem solving strategies

Strategy	Explanation
Trial and error	Continue trying different solutions until problem is solved
Algorithm	Step-by-step problem-solving formula
Heuristic	General problem-solving framework based on past experiences

Every problem is a feature space of all possible (successful or unsuccessful) solutions. The trick is to find an efficient search strategy. Simulation is one of the effective methods to solve problems.

3.1.2 System Modelling, Design, and Analysis

The difference between modelling, design, and analysis can be shown by the below explanation (Fig. 3.2).

Modeling is the process of gaining a deeper understanding of a system through imitation. Models specify **what** a system does. FSM, graphs, and mathematical models can be used for modeling (Fig. 3.3).

Design is the structured creation of artifacts. It specifies **how** a system does what it does. This includes optimization. Programming languages can be used for design.

Analysis is the process of gaining a deeper understanding of a system through dissection. It specifies **why** a system does what it does or fails to do what a model says it should do. **Simulation** methods can be used for analysis.

Fig. 3.2 Modelling, design, and analysis

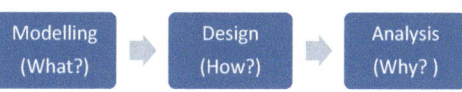

Fig. 3.3 A model has input X
and output Y

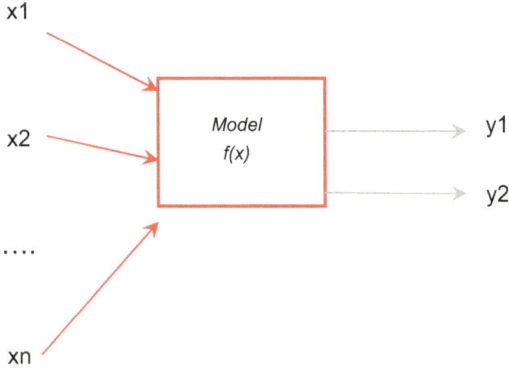

3.2   Anatomy of EDA Tools

EDA (Electronic Design Automation) tools are software tools used by engineers to design, simulate, and analyze electronic systems. The anatomy of EDA tools can be broken down into the following components where there are two primary categories: computer-aided design (CAD) tools and technology computer-aided design (TCAD) tools. The anatomy of these tools can be broken down as follows [3–5]:

1 CAD Tools: CAD tools are used for the design and layout of electronic systems. The anatomy of CAD tools includes the following components according to the functionality:
   - **Schematic capture**: This is the first step in designing an electronic system. EDA tools provide a graphical interface for engineers to create and edit circuit diagrams, which are used to represent the electronic components and their connections.
   - **Simulation**: EDA tools allow engineers to simulate the behavior of the electronic system under different conditions. This enables them to test and refine their designs before they are implemented in hardware.
   - **Layout**: Once the circuit design has been finalized, the next step is to create a physical layout of the components on a printed circuit board (PCB). EDA tools provide a user-friendly interface for engineers to arrange the components, define the routing of connections, and optimize the layout for performance and manufacturability.
   - **Design rule checking (DRC)**: DRC is a process that ensures the layout adheres to specific rules and constraints. EDA tools check the layout against a set of predefined design rules, which are based on the manufacturing process and the physical properties of the components. This helps to prevent errors and improve the reliability of the design.

- **Analysis**: EDA tools provide various analysis tools to help engineers evaluate the performance of the electronic system. This includes signal integrity analysis, power analysis, and thermal analysis. These tools help engineers identify potential problems and optimize the design for improved performance.
- **Verification**: Verification is a critical step in the design process that ensures the electronic system meets its functional and performance specifications. EDA tools provide various verification tools, such as functional simulation, timing analysis, and formal verification, to help engineers validate their designs.
- **Design management**: EDA tools provide a centralized platform for managing the design process, including version control, collaboration, and project management. This helps to streamline the design process and improve efficiency.

2 TCAD Tools: TCAD tools are used for the simulation and analysis of semiconductor devices. The anatomy of TCAD tools includes the following components:

- **Process simulation**: A tool for simulating the manufacturing process used to create semiconductor devices, including the deposition of materials, etching, and doping.
- **Device simulation**: A tool for simulating the behavior of semiconductor devices, such as transistors, diodes, and capacitors, under different operating conditions.
- **Modeling**: A tool for creating models of semiconductor devices that can be used in device simulation.
- **Analysis**: Various analysis tools, such as current–voltage (IV) analysis, capacitance–voltage (CV) analysis, and transient analysis.
- **Optimization**: A tool for optimizing the performance of semiconductor devices through parameter extraction and sensitivity analysis.

Anatomy of EDA Tools: CAD + TCAD is shown in Fig. 3.4 and a comparison is provided in Table 3.3. Technology Computer Aided Design (TCAD) tools are used for fabrication process, where it simulates the electrical characteristics of semiconductor devices. The EDA tools can be categorized according to the functionality:

- Design entry (capture tools)
- Synthesis tools
- Simulation tools
- IC physical design and layout tools
- IC verification tools
- PCB design and analysis tools

Simulators can also be categorized based on design level:

- Functional simulation
- Logic/Gate Level Simulation
- Switch/Transistor Level Simulation
- Circuit Simulation

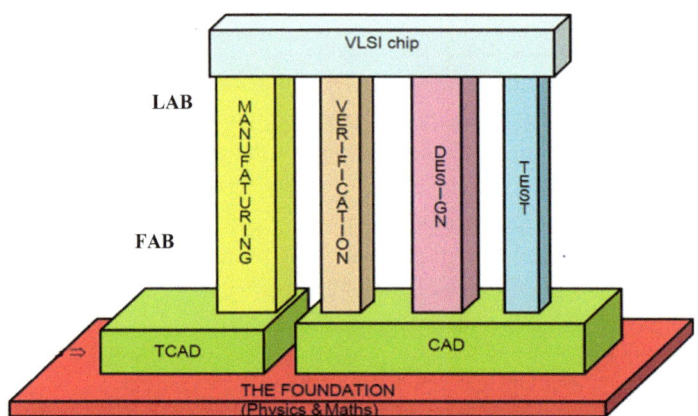

Fig. 3.4 Anatomy of EDA tools "EDA Space": CAD + TCAD. EDA is a technology enabler. **CAD + FAB + LAB**

Table 3.3 Comparison between CAD and TCAD tools

	CAD	TCAD
Definition	Computer-aided design	Technology computer-aided design
Focus	Design and layout of electronic devices	Simulation of electronic devices and circuits
Examples	Design of integrated circuits, PCBs, and MEMS	Simulation of device physics, TCAD process
Main goal	Optimize the design for performance and cost	Understand the behavior and performance of the device under different conditions
Input	Layouts and schematics	Physical and electrical parameters
Output	Schematics, layouts, and manufacturing files	Device characteristics and performance
Key features	Design rules, schematic capture, layout editor	Physical and electrical simulation, process simulation
Advantages	Faster design iterations, improved accuracy in layout, reduced errors	Improved device performance, reduced costs, better understanding of device behavior
Disadvantages	Limited understanding of device physics, does not consider physical limitations	Computationally intensive, requires expertise in device physics and simulation
Application	Used extensively in the semiconductor industry	Used for advanced research and development in the semiconductor industry

- Electromagnetic Simulation
- Device Simulation

Simulation is used to predict the circuit/system characteristic after manufacturing. Simulators allow you to test circuits without having to physically build them.

A High-Level communication system architecture is shown in Fig. 3.5 to explore different design alternatives: RF/Analog/Digital/Mixed.

RF (Radio Frequency) design involves the design and implementation of circuits and systems that operate at high frequencies, typically in the range of hundreds of MHz to several GHz. RF design is used in various applications, including wireless communication systems, radar systems, and satellite communication systems. Analog design involves the design and implementation of circuits that process continuous signals, such as audio or video signals. Analog design is used in various applications, including audio and video equipment, instrumentation, and control systems. Digital design involves the design and

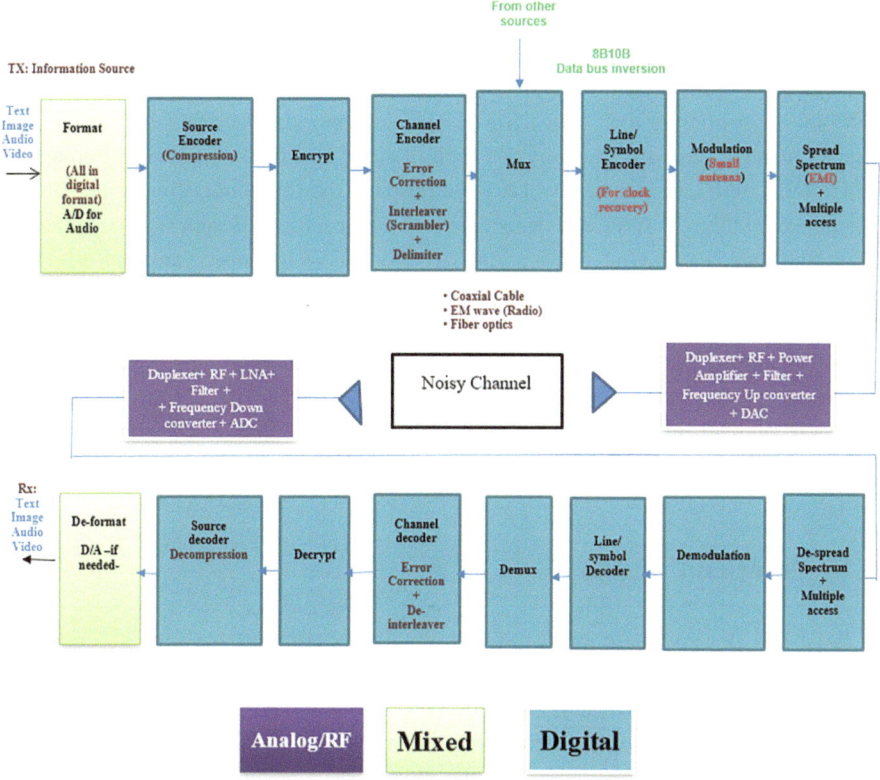

Fig. 3.5 A detailed block diagram of a digital communication system (PHY layer): Tx side, noisy channel, Rx Side (RF front end + baseband layer)

implementation of circuits that process digital signals, such as those used in computers and digital communication systems. Digital design is used in various applications, including computer hardware, embedded systems, and digital signal processing. Mixed-signal design involves the design and implementation of circuits that combine both analog and digital circuits on a single chip. Mixed-signal design is used in various applications, including data converters, communication systems, and control systems. Each of these design alternatives has its own strengths and weaknesses, and the choice of design approach depends on the specific application and the design requirements. For example, RF design is important for wireless communication systems, while digital design is important for computer hardware. Mixed-signal design is often used in applications that require both analog and digital circuits to work together seamlessly.

All EDA tools algorithms are based on the numerical analysis methods previously discussed in Chap. 2.

3.3   Anatomy of IC Design

The IC design flow is shown in Fig. 3.6. The first step in IC design is design specification (what customer wants) then we convert the specification to behavioral description. The behavioral description is then converted to RTL description. Then we perform functional verification and if there are any bugs, we fix it in the RTL then do the verification again. If the functional verification is ok, we start synthesizing the RTL code and do the gate level verification. By this, the front-end design is done. The back-end design starts by placement and routing then post-layout verification, we may repeat it if there are any errors until we generate the mask and send it to the fab. After fabrication, chip testing is done. There is a lot of SoC applications and corresponding IPs as shown in Table 3.4, where industry segments: including mobile communication, automotive, imaging, medical, and networking. The complete picture for electronic systems is described in Fig. 3.7. IC design is typically divided into two phases: front-end design and back-end design. Each phase involves different tasks and requires different skills.

3.3.1   Front-End Design

The front-end design phase of IC design involves designing the chip's functionality and behavior, typically using hardware description languages (HDLs) such as Verilog or VHDL. The front-end design includes tasks such as architecture definition, circuit design, and verification. During architecture definition, the system-level requirements are defined, and the overall architecture of the chip is decided upon. This includes identifying the required functional blocks and their interconnections. Once the architecture is defined, the circuit design process begins, where the individual circuits that make up the

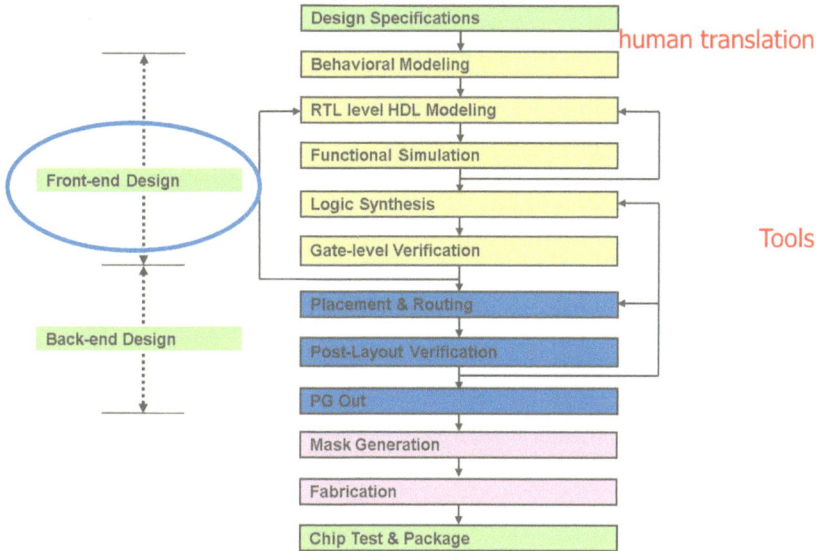

Fig. 3.6   A simplified high-level overview of IC design flow. PG stands for pattern generation

Table 3.4   SOC applications
and IPs examples

Category	IP
Processors	ARM
DSP	MPEG4,Viterbi
I/Os	PCI, USB
Mixed signal	ADC, DAC, PLL
Multimedia	HDMI
Memories	DRAM controller, flash memory
SoC buses	AHB
Miscellaneous	UART, Ethernet MAC

chip are designed. This includes designing logic gates, memory cells, and other circuit components. Once the circuit design is complete, the front-end design phase moves on to verification, which involves simulating the chip's behavior to ensure that it meets the functional and performance requirements. This process includes functional verification and performance verification. The result of front-end design is a netlist that represents the logical behavior of the design.

Fig. 3.7 Electronic systems
level from board to transistors

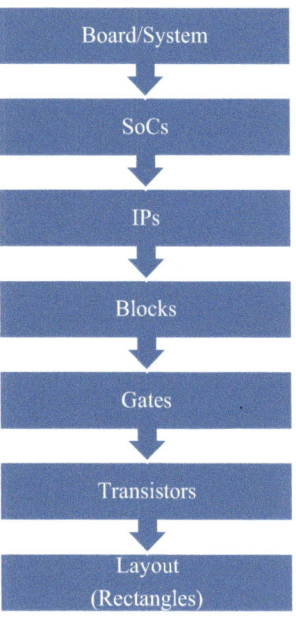

3.4   Back-End Design

The back-end design phase of IC design involves taking the netlist generated in the front-end design and converting it into a physical layout. The back-end design includes tasks such as floor planning, placement, routing, and physical verification. During floor planning, the physical layout of the chip is planned, and the size and location of the functional blocks are determined. Placement involves placing the logical elements of the design in specific locations on the chip. Once the placement is complete, routing begins, which involves connecting the logical elements with wires. This process includes both global and detailed routing. After the routing is complete, the chip is subjected to physical verification, which ensures that the design can be manufactured without defects. This includes Design Rule Checking (DRC), Layout versus Schematic (LVS) verification, and Electrical Rule Checking (ERC). The final result of the back-end design phase is the physical design database (PDB), which is used to manufacture the chip. In conclusion, front-end design is concerned with the logical functionality of the chip, while back-end design is concerned with the physical implementation of the design. Both front-end and back-end design are critical steps in the IC design process, and require different skills and expertise.

3.5   Frequency Spectrum

Electromagnetic waves consist of oscillating electric and magnetic fields, whose frequency (f) and wavelength (λ) are related as follows: $c = f\lambda$, where c is the speed of light, approximately 3×10^5 m/s. The spectrum is generally divided into seven regions in order of decreasing wavelength and increasing energy and frequency. The common designations are radio waves, microwaves, infrared (IR), visible light, ultraviolet (UV), X-rays and gamma-rays. The frequency range and its allocation are shown in Fig. 3.8. Electromagnetic waves are waves which can travel through the vacuum of outer space. Electromagnetic waves are created by the vibration of an electric charge. This vibration creates a wave which has both an electric and a magnetic component. In many applications, electromagnetic waves are generated and travel through free space at the speed of light. Radio waves and microwaves emitted by transmitting antennas are one form of electromagnetic energy. This electromagnetic energy is characterized by its frequency (in Hz) and wavelength. Radio frequency waves occupy the frequency range 3 kHz to 300 GHz. **Microwave**s are a specific category of radio waves that cover the frequency range 1 GHz to approximately 100 GHz (Table 3.5) [6].

Microwaves are produced by vacuum tubes devices that operate on the ballistic motion of electron controlled by magnetic or electric fields. Some different kinds of microwave emitters are the cavity magnetron, the klystron, the traveling-wave tube (TWT), the gyrotron and all stars. These devices work in the density modulated mode, instead of current modulated mode, meaning that they work based on clumps of electrons flying ballistically through them, instead of using a constant flow of electrons. Lower power microwaves can me produced by some solid-state devices such as the FET (field effect transistor), the tunnel diode, the Gunn diode, and the IMPATT diode.

Light has properties in common with both waves and particles. Light is the visible form of electromagnetic radiation.

RF and microwave radiation is **non-ionizing** because the energy levels associated with it are not high enough to cause ionization of atoms and molecules. Non-ionizing radiation does not have enough energy to knock electrons out of atoms. Near ultraviolet, visible light, infrared, microwave, radio waves, and low-frequency radio frequency (longwave) are all examples of non-ionizing radiation. By contrast, far ultraviolet light, X-rays, gamma-rays, and all particle radiation from radioactive decay are **ionizing**.

Light, radio waves, x-rays, ultra-violet radiation are all forms of a type of wave composed of oscillating electric and magnetic fields.

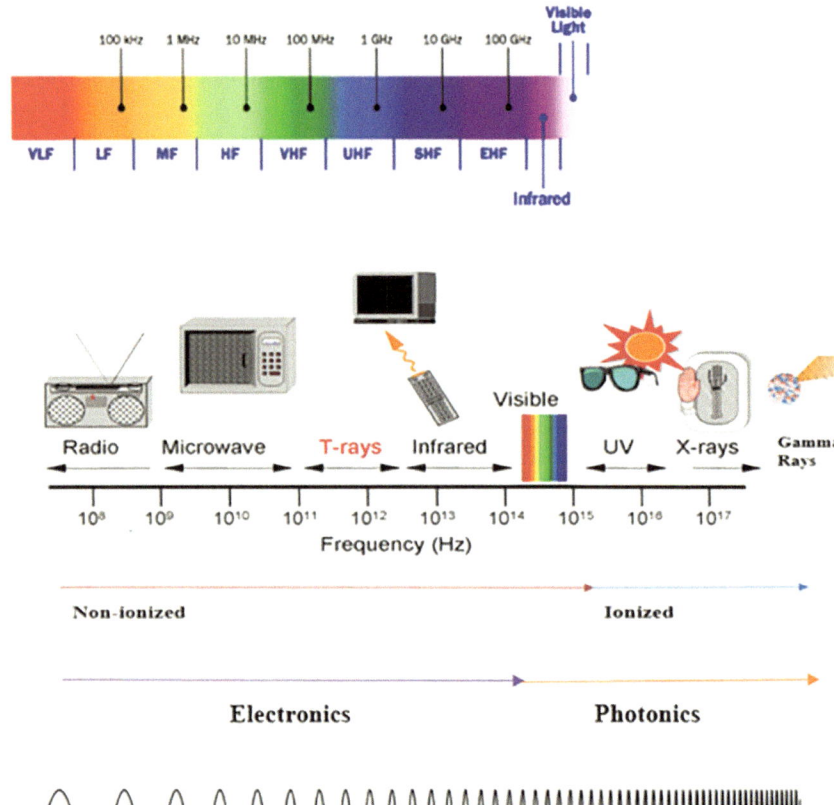

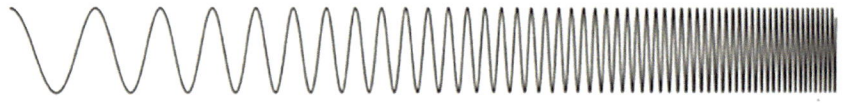

Fig. 3.8 The frequency range (logarithmic scale). T-rays is a terahertz band

Table 3.5 RF versus microwave: microwave is a subset of RF

Specifications	RF	Microwave
Frequency	3 kHz to 300 GHz	1–100 GHz
Applications	Mobile, TV	Radar, satellite, space
Source	Crystal oscillator	Tunnel diode

3.6   Classical Circuit Theory Versus Microwave Circuit Theory

A comparison between classical circuit theory and microwave circuit theory is shown in Table 3.6. Circuit simulation is used for Classical Circuit Theory while electromagnetic and device simulations are used for Microwave Circuit Theory (Table 3.7) [7, 8]. As per Fig. 3.8, for photonics, optical simulators are used such as Lumerical [9]. Lumerical develops high-performance photonic simulation software, enabling designers to predict light's behavior within complex structures and systems. Classical circuit theory and microwave circuit theory are two different approaches to analyzing and designing electrical circuits. Classical circuit theory is concerned with the analysis of circuits operating at frequencies below the microwave range. It focuses on the behavior of circuits in the time and frequency domains, using concepts such as voltage, current, resistance, capacitance, and inductance. It is primarily used for analyzing circuits in power electronics, signal processing, and control systems. Microwave circuit theory, on the other hand, deals with circuits that operate at high frequencies in the microwave range. This includes frequencies ranging from several hundred megahertz up to several gigahertz. Microwave circuits are used in a variety of applications, including communication systems, radar, and satellite systems. In microwave circuit theory, the physical characteristics of the circuit components, such as the dimensions of the transmission lines, are considered. One of the key differences between classical circuit theory and microwave circuit theory is the way in which the circuits are modeled. In classical circuit theory, circuits are typically modeled using lumped elements, which represent the electrical properties of components such as resistors, capacitors, and inductors. In microwave circuit theory, the circuit components are modeled using distributed elements, which consider the physical characteristics of the circuit components and the transmission lines. Another difference between the two approaches is the types of analysis that are performed. In classical circuit theory, circuit analysis is typically performed in the time or frequency domains. In microwave circuit theory, analysis is typically performed using S-parameters, which describe the relationship between the amplitude and phase of the signals at the input and output ports of the circuit.

3.7   Semiconductors Physics

Semiconductors are insulators when pure, a well-chosen impurity may make them conductors. Semiconductors can be grouped in two ways: direct bandgap semiconductor and indirect bandgap semiconductor. In direct bandgap semiconductor, the lower level of conduction band and higher level of valence band fall to the same level. In indirect bandgap semiconductor, electrons from conduction band fall to an intermediate state and from there to valence band. Energy will be released during this process in the form of heat. When electrons fall from conduction band to valence band energy is released which is

Table 3.6 Classical circuit theory versus microwave circuit theory

	Classical circuit theory	Microwave circuit theory	Quantum effect theory
Laws	Kirchhoff's laws	EM wave concept + classical circuit theory	Wave-particle theory
Used when	• Circuit dimension > 0.1 wavelength • For digital signal: pulse rise time) > 10 interconnect delay • For analog signal: carrier period > 10 interconnect delay time	Circuit dimension < 0.1 wavelength	If the transistor body dimension is at or below about 7 nm
Example			
Devices	Amplifier	Waveguide	Nanowires

equal to $E = hf$, where f: frequency, h: planks constant. If this frequency is in the range of visible light, then the semiconductor can emit light, the color of the light depends on the frequency which in turn depends on the bandgap. This is not the case with silicon and GaAs. That is why s they can't emit light. **Semiconductor devices** are light sensitive.

3.8   Why Is Simulation Needed?

Simulation is a powerful tool that plays a crucial role in various stages of product development, from conceptualization to manufacturing. It offers numerous benefits that contribute to efficiency, cost-effectiveness, and innovation across industries:

1. **Reduction of Physical Prototypes and Failure Prevention**
   - Simulation allows engineers and designers to create virtual models of products or systems, which can be tested under different conditions and scenarios without the need for physical prototypes.
   - By simulating various operating conditions and stress factors, potential failure points can be identified and addressed early in the design process, reducing the likelihood of costly errors and safety risks during product operation.
2. **Reduction of Development Time**
   - The ability to iterate rapidly through virtual simulations accelerates the development cycle by eliminating the time-consuming process of building and testing multiple physical prototypes.

Table 3.7 Comparison between circuit, electromagnetic simulator and device one [10–17]

Items	Circuit simulator	EM simulator	Device simulator	Quantum simulator	Optical simulator
Physics-based theory	Classical circuit theory	Electromagnetic-physics-based simulators	Semiconductors-physics-based simulators	Wave-particle theory	Snell's law
Equations	Kirchhoff's laws	Maxwell's equations (Maxwell's equations are essentially a mathematical description on how the electromagnetic fields are generated from sources ρ and J) where ε permittivity of the medium μ permeability of the medium σ conductivity of the medium E electric field H magnetic field ρ charge density J current density	The carrier transport equations (The motion of charged particles results in electric current: drift and diffusion) $J_n = \frac{\partial J_n(r,t)}{\partial t} + J_n(r,t) = q\mu_n n(r,t)E(r,t) + qD_n\nabla n(r,t)$ $\tau_p = \frac{\partial J_p(r,t)}{\partial t} + J_p(r,t) = q\mu_p p(r,t)E(r,t) + qD_p\nabla p(r,t)$ where τ_n average collision times of electron τ_n average collision times of holes J_n electron current density J_p hole current density μ_n effective carrier mobility of electrons μ_p effective carrier mobility of holes D_n diffusion coefficients of electrons D_p diffusion coefficients of holes n electron density p hole density E electric field q electron charge	• **Photoelectric effect**: electrons are emitted from atoms when they absorb energy from lights $E = hf$ E is the energy of electron h is Planck's constant f is photon frequency • **Schrodinger equation** $H\Psi = E\Psi$ H is the "Hamiltonian": a mathematical operator which represents the total energy, the sum of the kinetic energy and the potential energy Ψ is the so-called wavefunction	$n_1 \sin\theta_1 = n_2 \sin\theta_2$

(continued)

Table 3.7 (continued)

Items	Circuit simulator	EM simulator	Device simulator	Quantum simulator	Optical simulator
Disadvantages		• Does not consider particle theory of electromagnetic radiation (quantum effect), which is important in nanowires • Does not take into account the carrier transport equations	• Does not consider particle theory of electromagnetic radiation (quantum effect) • Does not consider magnetic field effects		
Simulators	SPICE	• HFSS [18], Q3D [19], HSPICE [20]	• SENTAURUS [21], SILVACO [22]		• COMSOL [23]

- Engineers can quickly evaluate design alternatives, make modifications, and optimize performance without waiting for physical components to be fabricated or assembled.

3. **Parameter Exploration and Predictive Maintenance**
   - Simulation enables engineers to explore the effects of different parameters on system behavior and performance.
   - By analyzing data generated from simulations, predictive maintenance strategies can be developed to anticipate equipment failures, optimize maintenance schedules, and minimize downtime.

4. **Simulation-Driven Experimentation and Manufacturing**
   - Simulation-based experimentation allows researchers and engineers to conduct virtual tests and experiments in a controlled environment.
   - This approach enables the optimization of manufacturing processes, material selection, and product design without the need for extensive physical testing.

5. **Design Space Exploration to Speed Up Analysis**
   - Simulation facilitates the exploration of a wide range of design possibilities within the constraints of performance, cost, and regulatory requirements.
   - By automating the analysis of simulation results, design space exploration tools help identify optimal solutions and refine design parameters more efficiently.
   - selection, and product design without the need for extensive physical testing.

6. **Engineering analysis versus simulation**

Engineering analysis and simulation are both integral parts of the product development process, but they serve distinct purposes and employ different methodologies. Here's a breakdown of each:

- **Engineering Analysis**
  - Engineering analysis involves the application of mathematical and scientific principles to assess the behavior, performance, and characteristics of a system or component.
  - It typically relies on theoretical models, equations, and empirical data to predict how a design will perform under specific conditions.
  - Engineering analysis may involve hand calculations, analytical methods, and simplified mathematical models to evaluate factors such as stress, strain, heat transfer, fluid dynamics, and structural integrity.
  - While engineering analysis provides valuable insights into the fundamental principles governing a system, it may have limitations in accurately predicting complex behaviors or interactions under real-world conditions.

- **Simulation**
  - Simulation involves the use of computer-based models to replicate the behavior and performance of a system in a virtual environment.
  - It encompasses various computational techniques, including finite element analysis (FEA), computational fluid dynamics (CFD), discrete event simulation, and multi-body dynamics.
  - Simulation allows engineers to explore a wide range of design parameters, operating conditions, and scenarios without the need for physical prototypes.
  - By iteratively refining the model and analyzing simulation results, engineers can optimize designs, identify potential issues, and make informed decisions to improve performance, reliability, and efficiency.
  - Simulation enables engineers to visualize complex phenomena, assess system behavior across different scales, and evaluate the impact of design changes in a cost-effective and timely manner.

3.9   CAD Simulation: RTL/Circuit/Electromagnetic/Device/ Optical/MEMS/Quantum/Acoustic/Thermal Simulation

Digital, analog, optical, RF, and Micro-Electro-Mechanical Systems (MEMS) are the foundations of different SoCs. Such SoCs will provide solutions in communication, networking, storage, computation, autonomous driving and more. Digital SoCs are designed to process digital signals and are commonly used in computing, communication, and networking applications. They use digital logic gates and circuits to process binary data and are capable of executing complex algorithms at high speed. Analog SoCs, on the other hand, are designed to process analog signals such as sound and video. They are used in audio, video, and sensor applications and require specialized circuits such as operational amplifiers, filters, and oscillators. Optical SoCs use light to transmit and process data. They are used in fiber-optic communication systems, sensors, and imaging applications. They typically use optical waveguides, photodetectors, and laser diodes to transmit and receive data. RF SoCs are designed to process radio-frequency signals and are used in wireless communication systems such as Wi-Fi, Bluetooth, and cellular networks. They use specialized circuits such as amplifiers, filters, and mixers to process and modulate RF signals. MEMS-based SoCs use micro-electro-mechanical systems to process and manipulate physical signals such as pressure, temperature, and motion. They are used in sensors, actuators, and biomedical devices and typically use microfabrication techniques to produce microscale components.

3.9.1   Logic/RTL Simulation

An RTL (Register Transfer Level) or logic simulator is a crucial tool in digital design for verifying and debugging the behavior of hardware designs described at the register transfer level. It operates by simulating the behavior of digital circuits specified in hardware description languages (HDLs) like Verilog or VHDL. RTL simulators translate the HDL code into a lower-level netlist, which represents the interconnections of basic logic gates and flip-flops. The simulator then executes the simulation by advancing time in discrete steps, evaluating the behavior of the design based on input stimuli and the current state of the system. Using **event-driven simulation**, RTL simulators propagate changes through the design, updating the state of registers and wires according to the logic and timing constraints specified in the HDL code. During simulation, designers can observe the behavior of signals over time through waveform traces and analyze simulation events and errors through log files. RTL simulators play a vital role in the digital design process, enabling designers to verify the correctness and functionality of their designs before synthesis and implementation on hardware platforms. They facilitate rapid design iterations, allowing designers to debug and refine their designs efficiently while ensuring that they meet functional requirements and performance constraints [24].

Cycle-accurate simulators represent a widely employed technique in computer-based simulations. They function by meticulously simulating systems in a clock-by-clock manner, capturing the intricacies of the system's internal operations. Specifically, these simulators emulate a system's micro-architecture by breaking down the simulation into discrete time-ticks that correspond to the system's clock speed. This approach finds significant utility in the design process of new microprocessors, allowing for the comprehensive simulation, testing, and debugging of the entire system, encompassing its operating system, compilers, and utilities, before physical chip manufacturing. Despite their precision at the clock cycle level, cycle-accurate simulators tend to operate at slower speeds compared to other simulation methods, sometimes exhibiting orders of magnitude difference. This limitation prompts researchers to explore various speed-up techniques like "reduced execution" or limiting the simulation to specific system components. However, adopting these strategies may compromise accuracy and result in errors. Although cycle-accurate simulators offer invaluable accuracy, they also present several disadvantages and constraints that hinder their widespread adoption across diverse applications. One major drawback is their inherent sluggishness compared to alternative simulators, which can impede practicality, especially when simulating large and complex systems. Additionally, cycle-accurate simulators often exist as closed-source systems, making it difficult to validate their accuracy and correctness against real hardware. Furthermore, the versatility and configurability of these simulators inversely correlate with their accuracy and reliability, rendering them less suitable for many real-world applications [25].

3.9.2   Circuit Simulation

Circuit simulation is a process of analyzing the behavior of an electronic circuit using software tools. It involves creating a virtual model of a circuit and simulating its operation under different conditions. The primary goal of circuit simulation is to predict the performance of a circuit before it is built, thus saving time, and reducing costs associated with physical testing. Circuit simulation tools use mathematical models to simulate the operation of a circuit. These models are based on the principles of circuit theory, which includes Ohm's Law, Kirchhoff's Laws, and other fundamental concepts. Circuit simulation software enables designers to create a circuit schematic and assign appropriate values to circuit components such as resistors, capacitors, inductors, and transistors. Once the circuit is created, simulation software can apply different input signals to the circuit and analyze its behavior, producing a graphical output that represents the circuit's response to different input conditions. Circuit simulation is essential in the design of electronic systems, particularly in complex circuits. It helps to identify potential problems and optimize circuit performance. Circuit simulation can be used in the design of various electronic systems, including analog circuits, digital circuits, power electronics, and radio-frequency (RF) circuits. It is also an essential tool in the design of integrated circuits (ICs) and field-programmable gate arrays (FPGAs).

3.9.2.1   Introduction: Digital/Analog/Mixed-Signal/RF

Logic circuits are built with transistors. Transistors is the smallest building block or device. Transistor operates as a simple switch. The most popular type of transistor for implementing a simple switch is the **Complementary Metal Oxide Semiconductor (CMOS).** In CMOS there are two types of transistors, PMOS and NMOS and together they **complement** each other.

A few typical functions will always remain analog such as Interfaces between the digital chip and the real world (ADC), Power management components (power amplifiers), Clock generation circuits (crystal oscillator), RF components in communication systems.

To design a logic circuit, several CAD tools are needed (Fig. 3.9). These tools can implement the following tasks: design entry, synthesis, physical design, simulation, and optimization. The first step is entering into the CAD system a description of the circuit/ system being designed. This can be done using **schematic capture** where logic circuit is defined by drawing logic gates and interconnecting them with wires or writing source code in a Hardware Description Language (HDL). The schematic capture tool provides a collection of graphical symbols that represent gates of various types with different number of inputs (library). An example is shown in Fig. 3.10. **Synthesis** is the process of generating a logic circuit from an initial specification that may be given in the form of a schematic diagram or code written in HDL and its behaviour can also be evaluated by means of simulation. **Functional simulator** uses the logic expressions (equations) generated during synthesis and assumes that these expressions will be implemented with perfect gates through which signals propagate instantaneously to verify that it will function as expected. The results of simulation are usually provided in the form of a timing diagram which the user can examine to verify that the circuit operates as required. For **physical design**, there are several different technologies that may be used to implement logic circuits. The physical design tools map a circuit specified in the form of logic expressions into a realization that makes use of the resources available on the target chip (Fig. 3.11). A **timing simulator** evaluates the expected delays of a designed logic circuit. Its results can be used to determine if the generated circuit meets the timing requirements of the specification for the design. **Optimization** is performed on the gate-level netlist for speed or for area. At this stage design can be simulated.

Fig. 3.9 From design entry to
physical design

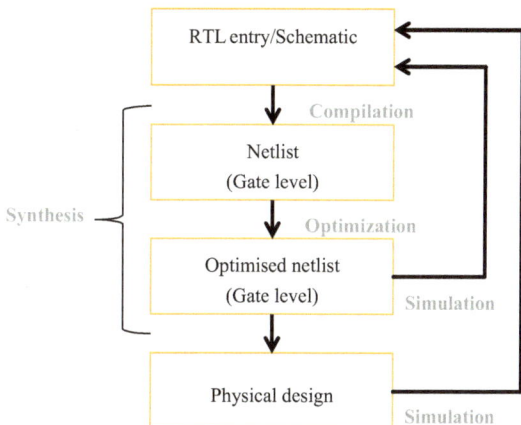

Fig. 3.10 **a** Schematic circuit
entry example, **b** the circuit
representation

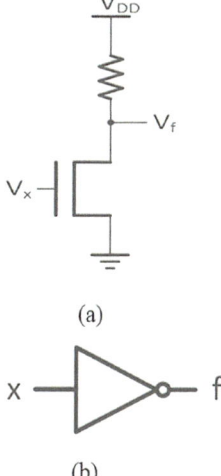

Fig. 3.11 **L** is usually set to the minimum value that is permitted according to process rules. Value of **W** is chosen depending on the amount of current flow, hence propagation delay, that is desired

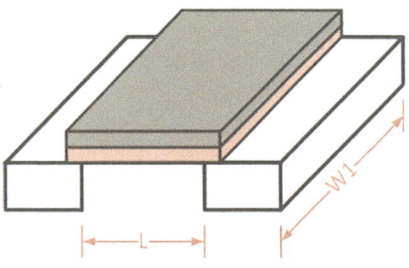

Fig. 3.12 An example of mixed-signal design

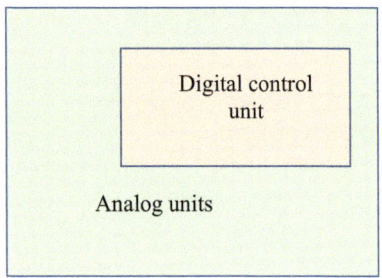

RF IC design is completely analog domain. **S-Parameter model** generation is an important aspect in RF simulation. **Mixed signals** contain both analog and digital. The Mixed Signal schematic is a conventional analog layout where a digital block is instantiated. After a digital cell described with RTL is imported into an analog library, three views are created. An example of mixed-signal design is shown in (Fig. 3.12). The design flow is shown in Fig. 3.13.

3.9.3   Electromagnetic Simulation

Electromagnetic simulation is a computational technique used to analyze and predict the behavior of electromagnetic fields in various environments. This technique involves the use of specialized software programs that can solve the mathematical equations governing electromagnetic fields, allowing engineers to study the behavior of electromagnetic waves in complex structures and environments. Electromagnetic simulation involves the use of numerical methods to solve Maxwell's equations, which describe the behavior of electromagnetic fields. These equations describe the relationships between electric and magnetic fields, and their interactions with charged particles and materials. Numerical methods, such as finite-difference time-domain (FDTD), finite-element method (FEM), and boundary element method (BEM), can be used to discretize the equations and solve them numerically. Electromagnetic simulation is used in a wide range of applications, such as antenna design, microwave and radio-frequency (RF) circuit design, electromagnetic

Fig. 3.13 Mixed-signal design flow. Circuit simulations include transient, DC sweep, AC sweep

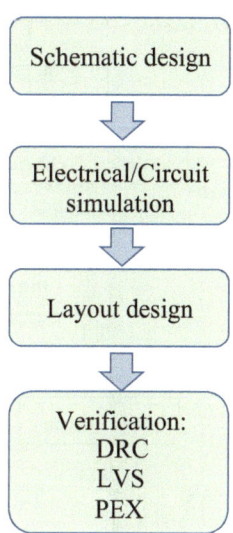

compatibility (EMC) analysis, electromagnetic interference (EMI) analysis, and electromagnetic scattering analysis. For example, in antenna design, electromagnetic simulation can help engineers optimize the performance of antennas by predicting their radiation patterns, impedance matching, and bandwidth. In RF and microwave circuit design, electromagnetic simulation can help to optimize the performance of filters, amplifiers, and other components. Electromagnetic simulation can also be used to analyze the electromagnetic compatibility of electronic systems, by predicting the levels of interference generated by the system and the susceptibility of the system to external electromagnetic fields. This information can be used to design and optimize shielding and grounding schemes, and to ensure compliance with regulatory standards. Several numerical schemes are used to discretize electromagnetics problems and solve Maxwell's equations in arbitrary geometries. The modeling methods can be categorized as: analytical, numerical, and empirical techniques (Fig. 3.14). The analytical techniques are equation-based one, while the empirical ones are measurement-based. Regarding numerical techniques, they are EM-simulation based and classified into integral-equation based and differential equation based [1]. Both of them can be represented in time domain or in frequency domain (Table 3.8), where PEEC is partial element equivalent circuit, MoM is method of momentum, FEM is finite element method, BEM is boundary element method, and FDTD is finite difference time domain. There is no exact analytical solution for many problems, that is why we need simulation.

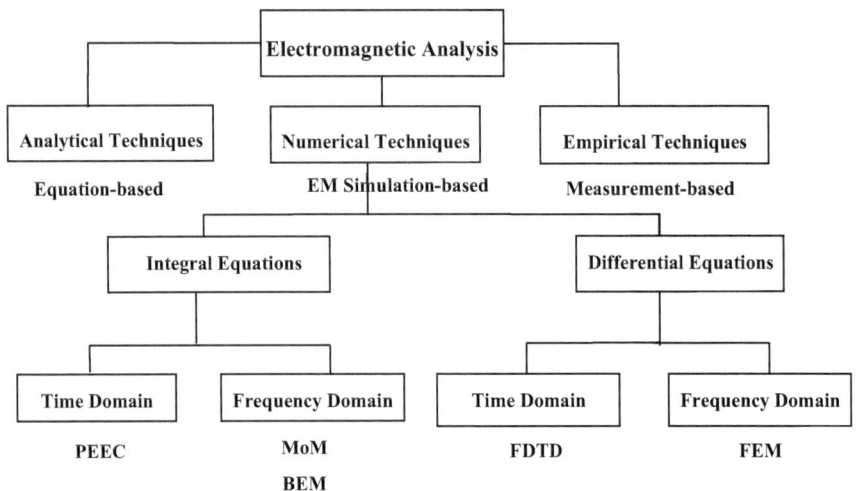

Fig. 3.14 Methods for solving Maxwell's equations

Table 3.8 Category of EM-simulation based techniques

	Time domain	Frequency domain
Integral equation based	PEEC	MoM, BEM
Differential equation based	FDTD	FEM

3.9.4 Device Simulation

Device simulation refers to the process of simulating the behavior and performance of individual semiconductor devices such as diodes, transistors, and integrated circuits (ICs). It involves modeling the physical and electrical properties of the device, and simulating its operation under different conditions, such as varying voltage, current, and temperature. Device simulation is an important tool in semiconductor device design and optimization, as it allows engineers to evaluate and optimize device performance before fabricating a physical prototype. It can also help identify potential failure modes and design limitations and provide insight into ways to improve device performance. Device simulation can be performed using a variety of software tools, including finite element analysis (FEA) and finite difference methods (FDM), as well as specialized software packages that are tailored to specific device types and applications. Overall, device simulation is a crucial aspect of semiconductor device design and development, enabling engineers to predict device behavior and optimize performance in a cost-effective and efficient manner.

Device simulator divides the simulated structure using an embedded meshing strategy taking into consideration the boundary conditions such as the electric field and the

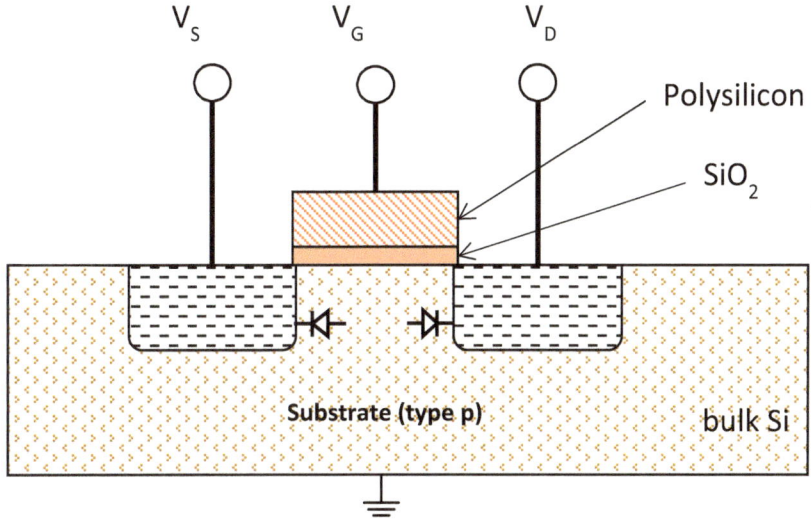

Fig. 3.15   Device representation for CMOS transistor

carrier currents. After meshing, it applies semiconductor equations and boundary condition on the meshed device structure by solving Poisson's equation and electron and hole continuity equations. While a purely electromagnetic (EM) approach, modeling the semiconductor as a constant conductivity medium, gives accurate results in the analysis of metal–semiconductor transmission line (e.g., it captures the slow-wave effect), this is not appropriate at all in describing the influence of DC bias devices. So, it is needed also to use device modeling approach [15, 26, 27]. The electromagnetic simulators are classified into three types of static, quasi static, and full wave. The main difference in the static simulator is that both skin effect and displacement current are neglected while in the quasi-static case, only the displacement current is neglected, and the full-wave simulator is the most accurate one. The electromagnetic simulator does not take into consideration the carrier transport equation, i.e., the nonlinearities of semiconductors, while the device simulator is semiconductors-physics-based one, but it does not consider magnetic field effects. Moreover, both simulators are neglecting quantum effect. An example is shown in Fig. 3.15.

3.9.5   Optical/Photonics Simulation

Instead of thinking about waves propagating through space, we can simply think about lines that are normal to the waves and we can call these lines light rays. The behavior of light rays can be modeled by some simple equations as Snell's Law. However, physical optics effects should be taken into consideration. **Wave-particle theory** explains how

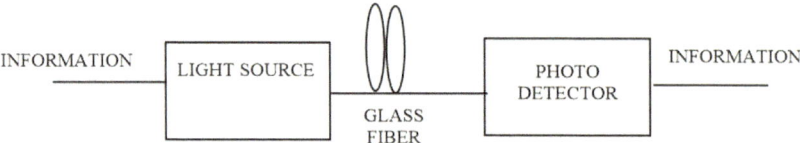

Fig. 3.16 Optical communication. Light source can be laser

electromagnetic radiation can behave as both a wave and a particle. Einstein argued that when an electron returns to a lower energy level and gives off electromagnetic energy, the energy is released as a discrete "packet" of energy. We now call such a packet of energy a **photon**. A photon resembles a particle but moves like a wave. In developing his quantum theory, Einstein suggested mathematically that electrons attached to atoms in a metal can absorb a specific quantity of light (first termed a **quantum**, but later changed to a **photon**) and thus have the energy to escape.

Laser beam can be transferred from the **generation medium** to the **target material** by two ways (Fig. 3.16):

1. **Optics**, such as driven mirrors reflect/deflect the laser beam onto the desired location on the target.
2. **Fiber waveguide**: Laser beam is coupled into the fiber or the flexible waveguide.

An optical system (Fig. 3.17) can be represented mathematically by:

$$\begin{bmatrix} y_2 \\ \theta_2 \end{bmatrix} = \begin{bmatrix} AB \\ CD \end{bmatrix} \begin{bmatrix} y_1 \\ \theta_1 \end{bmatrix} \tag{3.1}$$

Photonics means generation, transmission, and utilization of light. An example for silicon photonics is shown in Fig. 3.18.

Optical/Photonics simulation refers to the process of simulating the behavior and performance of optical components, devices, and systems. It involves modeling the physical and electromagnetic properties of light and simulating its interaction with various materials and structures. Examples of optical/photonics simulation include the design and

Fig. 3.17 Optical system

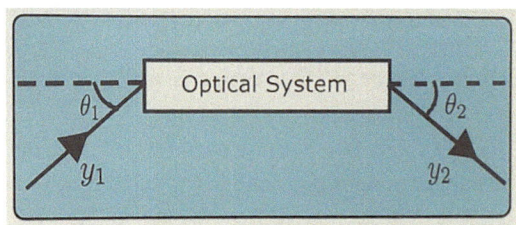

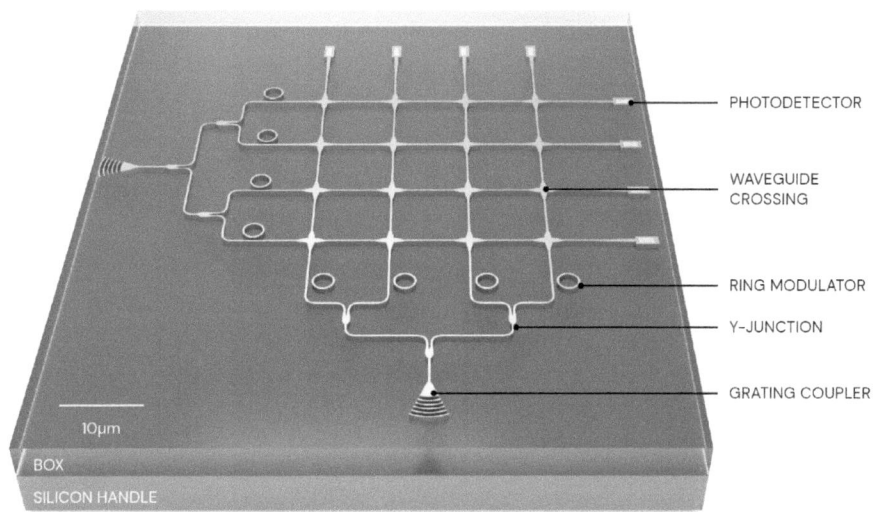

Fig. 3.18 Silicon photonics example

optimization of laser systems, fiber-optic communication networks, and imaging systems. For instance, in the design of a laser system, optical simulation tools can be used to model the behavior of light within the laser cavity, and optimize the laser's output power, spectral characteristics, and beam quality. In fiber-optic communication networks, optical simulation can be used to model the transmission of light signals through fiber-optic cables and optimize the network's performance in terms of signal quality, bandwidth, and distance. Optical simulation tools can also be used to design and optimize imaging systems, such as telescopes and microscopes, by modeling the behavior of light as it interacts with lenses, mirrors, and other optical components. Other examples of optical/photonics simulation include the design and optimization of photonic devices such as photovoltaic cells, optical modulators, and optical filters. In these applications, simulation tools can be used to model the behavior of light as it interacts with various materials, and optimize the device's performance in terms of efficiency, bandwidth, and spectral response.

3.9.6   MEMS Simulation

Physics such as electromechanics, piezo-electricity, piezo-resistivity, thermal-structure, and fluid–structure interactions can be modeled with Micro Electronic Mechanical Systems (MEMS) simulators. MEMS are used in many applications as shown in Table 3.9. MEMS simulation involves modeling the physical behavior of MEMS devices using simulation tools such as finite element analysis (FEA), computational fluid dynamics (CFD), and multi-physics simulation. The simulation process allows designers to predict the

Table 3.9 MEMS examples

Smart phones	• Microphones • RF filters • Motion sensors • Pressure sensors
Smart watches	• Motion sensors • Pressure sensors
Cars	• Microphones • Motion sensors • Pressure sensors • Gas sensors • LiDAR and infrared sensors for autonomous vehicles

performance of MEMS devices before manufacturing, optimize the device design, and improve the manufacturing process. Examples of MEMS simulation include the design and analysis of microcantilevers, microfluidic devices, micro-actuators, and microsensors. For instance, in microcantilever simulation, FEA tools can be used to model the deflection of a cantilever beam when a force is applied. This information can be used to design and optimize microcantilevers for use in various applications, such as biosensors for detecting biological molecules. Another example is in microfluidics simulation, where CFD tools can be used to model the flow of fluids through microchannels and optimize the design of microfluidic devices for applications such as lab-on-a-chip systems. MEMS simulation is becoming increasingly important as MEMS devices continue to be used in a growing number of applications, including biomedical sensors, telecommunications, and consumer electronics.

3.9.7   Quantum Simulation

These quantum effects become important in silicon **if the transistor body dimension is at or below about 7 nm**. As gate length is gradually reduced to accommodate scaling electrons can more easily **tunnel** from the source-drain. Certain semiconductor devices, under certain biasing condition (or in a certain mode), show the behavior that current increases when voltage decreases. Such devices are called **negative resistance** devices and are very commonly used in oscillators. The negative resistance device is the source of power. They are used in oscillators like tunneling diode. A tunnel diode is a type of semiconductor diode that has effectively "negative resistance" due to this quantum mechanical effect (tunneling). Tunnel diodes have a heavily doped positive-to-negative (P-N) junction that is about 10 nm wide. The heavy doping results in a broken band gap, where conduction band electron states on the N-side are more or less aligned with valence band hole states on the P-side. They are usually made from germanium, but can also be made from gallium arsenide and silicon materials. The tunnel diode showed great promise as an oscillator, Applications of tunnel diodes included local oscillators for UHF television tuners.

The Pauli Exclusion Principle states that no two fermions (a family of particles with half-integer spin, which includes electrons) may exist in the same energy state.

Heisenberg's Uncertainty Principle states that the more accurately you know the momentum or energy of a particle, the less accurately you will be able to know it's location in space or time. In other words, you can never determine position and velocity of a particle exactly.

The exponential growth in run-time for simulating a quantum computer with a classical machine prompts the need for an efficient simulator. The authors in [28] proposes a fast and flexible simulator for the evolution of the state of a quantum system by applying quantum gates to it. The authors choose Rust language [29] for implementing the simulator as implementing a fast quantum simulator requires the use of a programming language that provides blazingly fast performance, efficient use of memory, and a low level of control. The C and C++ programming languages fit the aforementioned criteria, but with many potential issues that are beyond the scope of this paper. The Rust programming language suffices all the aforementioned requirements for a fast quantum simulator, as well additional benefits. The primary benefit of Rust comes from the borrow checker, which helps eliminate memory safety bugs that continue to plague projects written in C and C++. Given that quantum computations are amenable to parallelism, Rust's type and ownership systems provide another benefit. Namely, concurrency errors that would only be caught at run-time in C and C++ are compile-time errors in Rust.

Quantum simulation is a method of using a quantum computer to simulate and study the behavior of quantum systems, which are difficult to study using classical computers. This approach involves mapping the quantum system of interest onto the quantum computer and then simulating the dynamics of the system. Quantum simulation is a promising area of research with potential applications in fields such as materials science, chemistry, and condensed matter physics. One of the key advantages of quantum simulation is its ability to efficiently solve problems that are intractable on classical computers. For example, simulating the behavior of a molecule with more than a handful of atoms is extremely challenging using classical methods, but can be accomplished using a quantum computer. This has important implications for drug discovery and materials design, as it allows scientists to simulate the behavior of complex molecules and materials to study their properties and interactions. Quantum simulation is an active area of research in both academia and industry, with many companies and research groups working to develop and improve the technology. Some of the current challenges in quantum simulation include developing better error correction methods to mitigate the effects of noise and decoherence, improving the scalability of the technology, and developing new algorithms and techniques to optimize the simulation process. Despite these challenges, quantum simulation holds great promise as a tool for exploring the behavior of complex quantum systems and driving advances in fields such as chemistry and materials science.

3.9.8 Acoustic Simulation

Acoustic simulation is a process of modeling and predicting the behavior of sound waves in a given environment. The simulation can be used to optimize acoustic designs for different applications, including architectural acoustics, noise control, and product design. Examples of acoustic simulation include:

- Room acoustics simulation: This involves modeling sound waves and their behavior in a given room or space. The simulation can be used to predict the acoustic characteristics of a space, such as its reverb time, sound pressure level, and speech intelligibility.
- Automotive acoustics simulation: This involves simulating the behavior of sound waves in a vehicle's interior and exterior. The simulation can be used to optimize the design of car interiors for improved noise reduction and better sound quality.
- Industrial noise control simulation: This involves modeling and predicting the behavior of sound waves in industrial environments, such as factories and power plants. The simulation can be used to design noise control measures, such as sound barriers and acoustic enclosures.
- Product design simulation: This involves simulating the acoustic characteristics of products, such as loudspeakers and headphones, to optimize their design for improved sound quality and performance.
- Aeroacoustics simulation: This involves simulating the behavior of sound waves in aerodynamic environments, such as aircraft engines and wind turbines. The simulation can be used to optimize the design of these systems for reduced noise emissions.
- Architectural acoustics simulation: This involves simulating the behavior of sound waves in large public spaces, such as concert halls and auditoriums. The simulation can be used to optimize the design of these spaces for optimal acoustic performance.
- Virtual reality acoustic simulation: This involves simulating the behavior of sound waves in a virtual environment, such as a video game or virtual training simulation. The simulation can be used to create immersive and realistic acoustic environments for a more engaging experience.
- Medical acoustics simulation: This involves simulating the behavior of sound waves in medical applications, such as ultrasound imaging and therapy. The simulation can be used to optimize the design of medical devices for improved accuracy and performance.
- Underwater acoustics simulation: This involves simulating the behavior of sound waves in underwater environments, such as sonar systems for navigation and communication.

The simulation can be used to optimize the design of these systems for improved performance and accuracy.
- Seismic acoustics simulation: This involves simulating the behavior of seismic waves, which are produced by earthquakes and other geological events. The simulation can be used to model and predict the effects of these events on buildings and other structures.

Acoustic simulation is important to minimize noise and optimize sound. FEM still can be used for simulating interior acoustics problems. Boundary element method (BEM) is used to solve exterior acoustics problems. The basic difference between these two techniques is the fact that BEM only needs to solve for unknowns on the boundaries, whereas FEM solves for unknowns in the volume. The governing equations are Newton's second law and continuity equation.

3.9.8.1 The 1D Acoustic Wave Equation Derivation
From Newton's Second Law

$$Force = mass \times acceleration \tag{3.2}$$

Dividing both sides of (3.2) by Volume, we get:

$$\frac{Force}{Volume} = \rho \times \frac{dv_x}{dt} \tag{3.3}$$

where $\rho = density\ of\ the\ medium$.
   Rewriting (3.3) it in terms of *Pressure Gradient*:

$$-\frac{dP}{dx} = \rho \times \frac{dv_x}{dt} \tag{3.4}$$

where $P = Pressure$, the negative sign in (3.4) accounts for the fact that Force due to a Pressure Gradient is in the direction of decreasing Pressure.
   Differentiating (3.4) w.r.t position:

$$-\frac{\partial^2 P}{\partial x^2} = \rho \frac{\partial}{\partial t}\left(\frac{\partial v_x}{\partial x}\right) \tag{3.5}$$

From the continuity equation
Consider a small section (length Δx) of a tube (cross section A). This tube is filled with a material that has Bulk Modulus B defined as:

$$\Delta P = -B\frac{\Delta V}{V} \tag{3.6}$$

The Volume of the small section is:

$$V = A(\Delta x) \tag{3.7}$$

Due to a disturbance the particles of the material move from their original position by a position dependent amount s(x). The change in Volume is:

$$\Delta V = A(\Delta s) \tag{3.8}$$

Substituting (3.7), (3.8) into (3.6):

$$\Delta P = -B\frac{A(\Delta s)}{A(\Delta x)} \tag{3.9}$$

Equation (3.9) can be represented as:

$$\partial P = -B\frac{\partial s}{\partial x} \tag{3.10}$$

Differentiating (3.10) with respect to time, we get:

$$\frac{\partial P}{\partial t} = -B\frac{\partial v_x}{\partial x} \tag{3.11}$$

Differentiating Continuity Eq. (3.11) w.r.t time:

$$\frac{\partial^2 P}{\partial t^2} = -B\frac{\partial}{\partial t}\left(\frac{\partial v_x}{\partial x}\right) \tag{3.12}$$

Combining (3.5) and (3.12), we get the Acoustic Wave Equation:

$$\frac{\partial^2 P}{\partial t^2} = \frac{B}{\rho} \times \frac{\partial^2 P}{\partial x^2} \tag{3.13}$$

Equation (3.13) can be represented as:

$$\frac{\partial^2 P}{\partial t^2} = c^2 \times \frac{\partial^2 P}{\partial x^2} \tag{3.14}$$

where $\sqrt{\frac{B}{\rho}} = c$, and c is the speed of sound.

3.9.8.2 The 3D Acoustic Wave Equation Derivation

We can extend to 3D by using gradients and divergence instead of spatial derivatives and substitute in (3.4) and in (3.14) which results in (3.15) and (3.16) respectively:

$$-\nabla P = \rho \times \frac{\partial v}{\partial t} \tag{3.15}$$

$$\frac{\partial P}{\partial t} = -B\nabla \cdot v \tag{3.16}$$

We get the 3D e Acoustic Wave Equation by combining (3.15) and (3.16) or substituting in (3.14):

$$\frac{\partial^2 P}{\partial t^2} = c^2 \times \nabla^2 P \tag{3.17}$$

where $\nabla^2 = \left(\frac{\partial^2}{\partial^2 x} + \frac{\partial^2}{\partial^2 y} + \frac{\partial^2}{\partial^2 z}\right)$ is the Laplacian.

Among the solutions to the Acoustic Wave Equation, of particular interest are those with sinusoidal time dependence:

$$P(x, t) = \mathrm{Re}\left(u(x)e^{-i\omega t}\right) \tag{3.18}$$

Substituting (3.18) into (3.17) results in:

$$\nabla^2 u + k^2 u = 0 \tag{3.19}$$

where $k = \frac{\omega}{c} = \frac{2\pi f}{c} = \frac{2\pi}{\lambda}$ is the wave number.

The Acoustic Wave Equation is a PDE in space *and* time. So, FEM can be used to solve it.

If there is a sound source, then (3.17) can be represented as:

$$\frac{\partial^2 P}{\partial t^2} - c^2 \times \nabla^2 P - f(x, t) = 0 \tag{3.20}$$

where f(x, t) represents sound sources.

3.9.9 Thermal Simulation

Thermal simulation refers to the process of simulating heat transfer and thermal behavior of a system or component. This is done by using computer-aided engineering (CAE) tools that allow engineers to model and simulate the thermal characteristics of a system and make predictions about its performance under different conditions. Examples of thermal simulation applications include:

- Electronic Cooling: Simulation of the thermal behavior of electronic components and devices, including printed circuit boards (PCBs), processors, and power supplies.
- Building Energy Analysis: Simulation of the thermal performance of buildings and their HVAC (heating, ventilation, and air conditioning) systems, in order to optimize energy efficiency and reduce costs.

- Automotive Cooling: Simulation of the thermal performance of automotive components and systems, such as engine cooling and air conditioning.
- Industrial Process Heating: Simulation of the thermal behavior of industrial furnaces, ovens, and other heating systems, in order to optimize performance and reduce energy consumption.
- Aerospace Heat Transfer: Simulation of the thermal behavior of aerospace components and systems, such as aircraft engines, spacecraft heat shields, and rocket nozzles.

Thermal simulation is critical for understanding the behavior of systems that involve heat transfer, and for optimizing their performance. By simulating the thermal behavior of a system, engineers can identify potential problems and make design changes to improve its efficiency and effectiveness. *Thermal analysis* is a group of techniques that looks at how the physical properties of materials change with changes in temperature. Thermal analysis simulation analyzes heat transfer processes and aids in their design. These processes are linear, non-linear, transient, and steady. The software provides solutions to methods, including conduction, convection, radiation, and changes of phase.

3.9.9.1 Heat Equation
The heat equation is given by [30]:

$$\alpha \nabla^2 u = \frac{\partial u}{\partial t} \qquad (3.21)$$

where u is the temperature, α is the thermal diffusivity constant. To solve such as equation we need to know the boundary conditions then apply FEM or any other method that was discussed previously to solve PDE.

3.10 TCAD Simulation: Process and Manufacturing/Fabrication Simulation

TCAD tools play important roles for the semiconductor device industry including in-depth understanding the underlying physics of device operation, device optimization reducing costly wafer reruns in the foundry, debugging fabrication faults, exploring, and isolating the impact of different process and statistical variability sources [31]. Process simulation such as lithography simulation is a critical step in VLSI design for manufacturability as in the advanced CMOS process, more timing failures are induced by process variations. **Process space exploration** is used for the design, development, analysis, and optimization of fabrication processes.

TCAD simulators can model a wide range of physical process behavior such as lithography, etching, deposition and patterning. Moreover, to develop process flows and perform experiments before actual fabrication. The growing complexity and shrinking geometries of recent manufacturing technologies are making high-density, low-voltage ICs increasingly susceptible to the influences of electrical noise, process variation, transistor aging, and the effects of natural radiation. Thus, TCAD simulation is inevitable.

3.10.1   Lithography Process

Lithography is a process used in microfabrication to pattern parts of a thin film or the bulk of a substrate. It uses light to transfer a geometric pattern from a photomask to a light-sensitive chemical "photoresist on the substrate. In the lithography process, a mask pattern is transferred to a photosensitive layer on a semiconductor wafer through a projection lens system. The TCAD simulation can be used to predict the intensity distribution of the light and the resulting pattern on the wafer. The simulation can also be used to optimize the parameters of the lithography process, such as the exposure dose and focus, to achieve the desired pattern quality. TCAD simulation can also be used to simulate the effect of process variations on the lithography process. For example, the simulation can be used to predict the effect of variations in the thickness of the photoresist layer or variations in the critical dimensions of the mask pattern on the resulting pattern on the wafer. Overall, TCAD simulation for lithography process can help semiconductor manufacturers optimize the lithography process and improve the yield and performance of their devices. Comparison between different types of lithography technology is shown in Table 3.10.

3.10.1.1   Electron-Beam Lithography
Electron-beam lithography (EBL) is the preferred patterning method for product development. In EBL, a resist layer is directly patterned by scanning with an electron beam

Table 3.10 Comparison between different types of lithography technology

	Electron-beam lithography	Proximity lithography	Laser lithography
Resolution	Highest	Moderate	Moderate to high
Exposure time	Long	Short	Short
Fabrication time	Long	Short	Short
Complexity	High	Low	Moderate to high
Cost	High	Low	Moderate to high
Mask requirement	No mask needed	Mask required	Mask required
Fabrication volume	Small	Large	Moderate
Applicability	Research and small-scale manufacturing	Mass production and large-scale manufacturing	Mass production and large-scale manufacturing

electronically. EBL also has the advantage of allowing multiple designs to be fabricated together on one wafer. EBL is, however, a slow and expensive process, which is not practical for production. Substrate charging and proximity error effects must be considered to get good quality devices. Proximity error correction effects are overcome using specialized design correction software.

3.10.1.2 Proximity Lithography

Mask is placed in the direct vicinity of a wafer, eliminating the optical system otherwise needed for the projection of the mask image onto the wafer. Proximity lithography may offer a low-cost alternative to the traditional imaging approach with potentially high throughput.

3.10.1.3 Laser Lithography

Laser lithography is a versatile technique for the creation of microstructures such as microelectromechanical systems (MEMS) and integrated circuits. It is ideal for MEMS, BioMEMS, nanotechnology, and many more.

3.10.2 Patterning Processes

Patterning processes include exposure, development, etching, and ion implantation. **Etching process** is a step to remove the lower part of the layer not covered by the photoresist (PR) following the photo process with an aim to leave the necessary pattern only [32]. The patterning process involves the transfer of a pattern from a photomask onto a semiconductor wafer through the use of various lithography techniques. TCAD simulation can be used to analyze and optimize the patterning process and the quality of the resulting pattern. This simulation helps to identify process variations, predict process-induced defects, and improve yield and reliability. TCAD simulation for patterning processes involves modeling the behavior of light, photoresist, and other materials during the lithography process. This includes simulating the propagation of light through the lens system, diffraction effects, the chemical reaction of the photoresist with light, and the development process that removes the unwanted photoresist material. The simulation can also include the etching process used to transfer the pattern into the underlying layers of the device. The simulation provides information on the critical dimensions, overlay accuracy, and edge roughness of the final pattern, which is essential for device performance. TCAD simulation for patterning processes can also be used to optimize the lithography process parameters, such as exposure dose, focus, and mask design, to achieve the desired pattern quality and device performance. Overall, TCAD simulation for patterning processes is a powerful tool for optimizing the lithography process and improving device performance and yield.

3.10.3 Material Modeling

TCAD simulation for material modeling is a process of creating models of semiconductor materials and analyzing their electrical and thermal properties using simulation software. These models can then be used to optimize the design and performance of semiconductor devices. The process of TCAD simulation for material modeling involves creating a virtual structure of the material being studied using a computer-aided design (CAD) software. The properties of the material, such as its crystal structure, doping concentration, and temperature, are then defined. The simulation software then uses mathematical models to simulate the behavior of electrons and holes in the material under various conditions, including the application of an electric field or temperature changes. TCAD simulation for material modeling can be used to study a wide range of semiconductor materials, including silicon, gallium arsenide, and indium phosphide. It can also be used to simulate the behavior of different doping elements, such as boron, phosphorus, and arsenic. The results of TCAD simulation for material modeling can provide valuable insights into the behavior of semiconductor materials and help optimize the design and performance of semiconductor devices. For example, it can be used to study the impact of different doping levels and material properties on the performance of a transistor or other semiconductor device. Recent advances in TCAD simulation for material modeling have focused on improving the accuracy and speed of simulations, as well as incorporating more complex physical models into the simulation software. Material modeling is **an equation that relates the applied force (or stress) to the resulting deformation (or strain)**. All FEM solvers support many material models, such as linear elastic, elastic–plastic, etc. [32].

3.10.4 Plasma Simulation

Plasma simulation is a computational method used to model the behavior of plasma, which is a state of matter consisting of ionized gas. Plasma simulation is crucial in various fields such as astrophysics, fusion research, space physics, and semiconductor manufacturing. It involves complex mathematical models and numerical techniques to understand plasma dynamics, including particle interactions, electromagnetic fields, and plasma instabilities. Scientists and engineers use plasma simulation to predict and analyze phenomena, optimize experimental setups, and design future technologies like fusion reactors and plasma thrusters. Various software packages and simulation codes are available to perform plasma simulations, each tailored to specific applications and research goals.

References

1. Han, K. J. (2009). *Electromagnetic modeling of interconnections in three-dimensional integration.* PhD Dissertation, Georgia Institute of Technology.
2. https://opentext.wsu.edu/psych105/chapter/7-4-problem-solving
3. Paknikar, R., Bansode, S., Nandihal, G., Desai, M. P., Moudgalya, K. M., & Jha, A. (2021). eSim: An open source EDA tool for mixed-signal and microcontroller simulations. In *2021 4th International Conference on Circuits, Systems and Simulation (ICCSS)*, Kuala Lumpur, Malaysia (pp. 212–217). https://doi.org/10.1109/ICCSS51193.2021.9464198
4. Dilip Save, Y., et al. (2013). Oscad: An open source EDA tool for circuit design, simulation, analysis and PCB design. In *2013 IEEE 20th International Conference on Electronics, Circuits, and Systems (ICECS)*, Abu Dhabi, United Arab Emirates (pp. 851–854). https://doi.org/10.1109/ICECS.2013.6815548
5. Lück, C., Lopera, D. S., Wenzek, S., & Ecker, W. (2022). Industrial experience with open-source EDA tools. In *2022 ACM/IEEE 4th Workshop on Machine Learning for CAD (MLCAD)*, UT, USA, 2022 (pp 143–143). https://doi.org/10.1109/MLCAD55463.2022.9900097
6. Mohamed, K. S. (2022). wireless communication systems: Foundation. In *Wireless communications systems architecture*. Synthesis lectures on engineering, science, and technology. Springer. https://doi.org/10.1007/978-3-031-19297-5_1
7. Salah, K., Ismail, Y., El-Rouby, A., Salah, K., Ismail, Y., & El-Rouby, A. (2015). 3D/TSV-enabling technologies. *Arbitrary modeling of TSVs for 3D integrated circuits* (pp. 17–47).
8. Salah, K., et al. (2015). TSV fabrication. *Arbitrary modeling of TSVs for 3D integrated circuits* (pp. 163–172).
9. https://www.lumerical.com/
10. Ho, W., Yoon, S.W., Zhou, Q., Pasad, K., Kripesh, V., & Lau, J. H. (208). High RF performance TSV silicon carrier for high frequency application. In *Electronic Components and Technology Conference*.
11. Beece, A., Rose, K., Zhang, T., & Qiang, J. (2009). Modeling and evaluation for electrical characteristics of Through-Strata-Vias (TSVs) in three-dimensional integration. In *3D System Integration, 2009. 3DIC 2009. IEEE International Conference*.
12. Shim, J., Song, E., Pak, J., Lee, J., Lee, H., Park, K., &Kim, J. (2009). Active circuit to through silicon via (TSV) noise coupling. In *Electrical Performance of Electronic Packaging and Systems, 2009. EPEPS '09. IEEE 18th Conference*.
13. Kim, J., Song, E., Cho, J., Pak, J., Lee, J., Lee, H., Park, K., & Kim, J. (2009). Through silicon via (TSV) equalizer. In *Electrical Performance of Electronic Packaging and Systems, 2009. EPEPS '09. IEEE 18th Conference*.
14. Rousseau, M., Rozeau, O., Cibrario, G., Le Carval, G., Jaud, M.-A., Leduc, P., Farcy, A., & Marty, A. (2008). Through-silicon via based 3D IC technology: Electrostatic simulations for design methodology. In *IMAPS Device Packaging Conference*, Phoenix, AZ, United States.
15. Bertazzi, F., Cappelluti, F., Bonani, F., & Ghione, G. (2009). *Accurate simulation of travelling-wave electro-absorption modulators through a novel coupled electromagnetic and carrier-transport model*. IEEE.
16. Salah, K., et al. (2012). Modeling and analysis of through silicon via: Electromagnetic and device simulation approach. In *2012 19th IEEE International Conference on Electronics, Circuits, and Systems (ICECS 2012)*. IEEE.
17. Salah, K., & Ismail, Y. (2015). Design of adiabatic TSV, SWCNT TSV, and Air-Gap Coaxial TSV. In *2015 IEEE International Symposium on Circuits and Systems (ISCAS)*.
18. http://www.ansoft.com/products/si/hfss

19. http://www.ansoft.com/products/si/q3-dextractor
20. Star-Hspice Reference Manual, Avant! Corporation, January 1999.
21. Sentaurus Device User Guide, Ver. A-2008.09, Synopsis Inc.
22. http://www.silvaco.com/
23. https://www.comsol.co.in.
24. Kupriyanov, A., Hannig, F., & Teich, J. (2004). High-speed event-driven RTL compiled simulation. In *International Workshop on Embedded Computer Systems*. Springer.
25. Schreiner, S., Görgen, R., Grüttner, K., & Nebel, W. (2016) A quasi-cycle accurate timing model for binary translation based instruction set simulators. In *2016 International Conference on Embedded Computer Systems: Architectures, Modeling and Simulation (SAMOS)*, July 2016 (pp. 348–353).
26. Wang, G. (2001). *Coupled electromagnetic and device level investigations metal–insulator–semiconductor interconnects.* Ph.D. dissertation Dept. Sci. Comput., Stanford Univ., Stanford, CA, 2001.
27. Sze, S. M., & Ng, K. (2006). *Physics of semiconductor.* Wiley.
28. Yusufov, S., Stefanski, C., & Gonciulea, C. (2023). Designing a fast and flexible quantum state simulator. arXiv preprint arXiv:2303.01493
29. Klabnik, S., & Nichols, C. (2019). *The rust programming language.* No Starch Press. ISBN: 9781718500440.
30. Cole, K. D., Beck, J. V., Haji-Sheikh, A., & Litkouhi, B. (2011). *Heat conduction using green's functions* (2nd ed.). CRC Press.
31. Medina-Bailon, C., et al. (2020). *Journal of Microelectronic Manufacturing, 3*(4), 20030404.
32. https://news.skhynix.com/etching-process-to-complete-semiconductor-patterning-1/

4.1 Introduction

Machine learning can be used to enhance chip design simulation. ASIC design flow is shown in Fig. 4.1. AI/ML can be applied to each step. In this chapter, we will discuss how ML can be deployed to enhance ASIC/FPGA design flow from EDA/simulation perspective and accelerate design and verification steps. The emergence of new technologies with unconventional requirements and Saturation of Moore's law due to limitations of CMOS technology are motivators for deploying AI in EDA tools as recent advances in deep learning makes it a possible deployment. Deep learning stacks many hidden layers together, allowing for more sophisticated processing of training data. Moreover, it can extract the relevant features of any problem domain as at every layer, DL models use shift-invariant filters to perform convolution, followed by a non-linear operation such as rectified-linearity (ReLU). Any ML algorithm should be autonomous and generic (Table 4.1).

Machine learning for building the next generation of EDA tools can be used for parameter tuning and feature detection or pattern recognition. ML-assisted simulators combine traditional simulators with machine learning techniques to improve simulation speed and accuracy. These simulators use machine learning algorithms to learn and predict the behavior of a system under simulation, and to optimize the simulation parameters. Traditional simulators rely on analytical models and numerical methods to simulate the behavior of a system. However, these simulations can be computationally expensive, especially for complex systems. ML-assisted simulators address this issue by using machine learning algorithms to accelerate the simulation process. One common approach in ML-assisted simulators is to use surrogate models, which are machine learning models trained on a limited number of simulations runs to predict the behavior of the system for a large number of simulation scenarios. These surrogate models can be much faster to evaluate

© The Author(s), under exclusive license to Springer Nature Switzerland AG 2025   131
K. S. Mohamed, *Next Generation EDA Flow*, Synthesis Lectures on Engineering,
Science, and Technology, https://doi.org/10.1007/978-3-031-88435-1_4

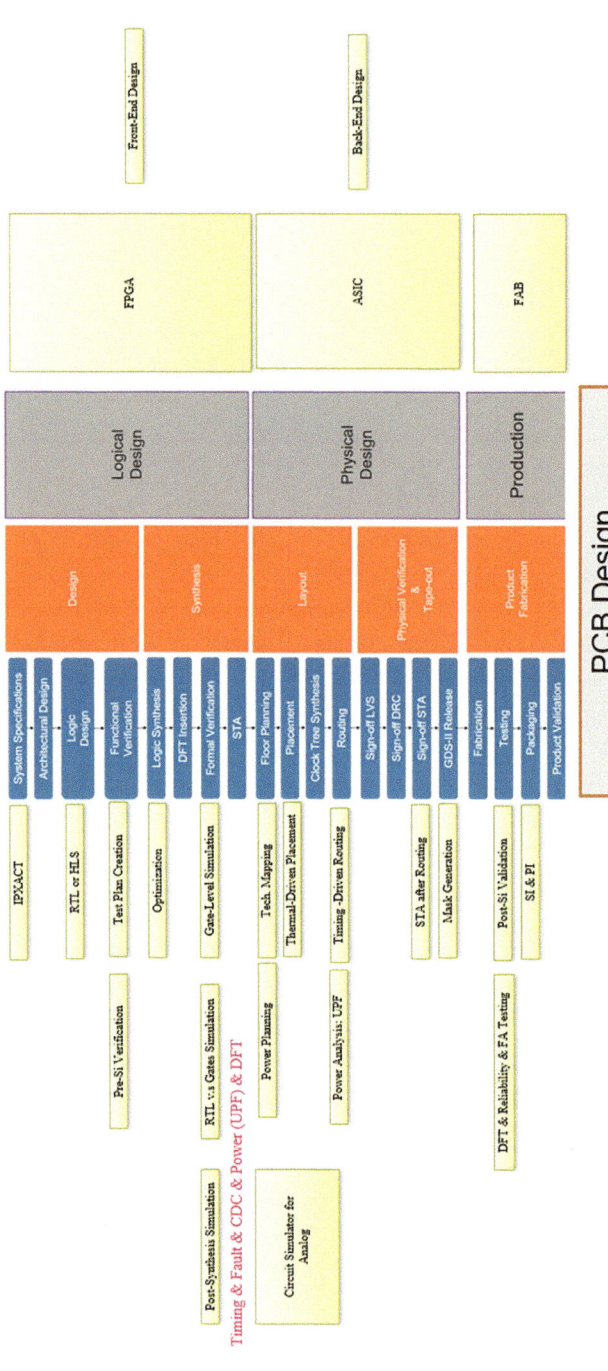

Fig. 4.1 ASIC design flow. It has long and complicated design flow and requires many iterations for convergence. Single iteration is expensive. So, ML can reduce this expensive iterative process. ML can be used to **enhance existing EDA Tools or to propose Novel EDA Applications**

Table 4.1 Properties of ML algorithm

Property	Description
Accuracy	Ability to provide outputs that are close to true values
Interpretability	Ability to explain reasoning behind outputs in a human-understandable way
Scalability	Ability to handle large amounts of data and scale to larger systems
Robustness	Ability to function correctly even with variations in data or different environments
Generalization	Ability to perform well on new, unseen data
Efficiency	Ability to perform tasks in a timely manner and use resources effectively
Privacy	Ability to preserve the privacy of the data being processed
Security	Ability to protect against unauthorized access, modification, or destruction of data
Autonomous	It learns automatically and does not require human intervention
Generic	The same algorithm can be used to achieve different optimization objectives
Transferability	Ability to apply learned knowledge from one domain to another domain

than the full simulator and can be used to explore the behavior of the system over a wide range of parameters. Another approach is to use machine learning algorithms to optimize the simulation parameters. For example, reinforcement learning can be used to find the optimal parameter settings that minimize the simulation error or maximize some other performance metric. This can significantly reduce the time required to perform simulations and improve the accuracy of the results. Overall, ML-assisted simulators can provide a powerful tool for designing and optimizing complex systems, especially when traditional simulation methods are computationally expensive or infeasible [15].

The next generation Electronic Design Automation (EDA) flow is a new approach to designing integrated circuits that aims to overcome the limitations of traditional EDA flows. The traditional EDA flow is a sequential process that involves multiple steps, such as logic design, synthesis, placement and routing, and verification, with each step being performed independently. In contrast, the next generation EDA flow is a more holistic approach that integrates different design steps and uses advanced algorithms and technologies to optimize the design process. Some key features of the next generation EDA flow include:

- **Machine learning**: Machine learning algorithms can be used to optimize the design process, such as predicting the performance of the circuit, generating layouts, and reducing power consumption [14].

- **Cloud computing**: Cloud computing can provide the necessary computing resources for running complex simulations and optimizing the design process, enabling faster and more efficient design cycles.
- **High-level synthesis**: High-level synthesis tools can automatically generate optimized hardware designs from high-level descriptions, reducing the need for manual coding and improving productivity.
- **Advanced verification**: Advanced verification techniques, such as formal verification and emulation, can ensure the correctness of the design and reduce the risk of design errors.
- **Design for manufacturability**: The next generation EDA flow incorporates design for manufacturability (DFM) considerations early in the design process, ensuring that the final design is manufacturable with high yield.
- **Design for reliability**: The next generation EDA flow also considers reliability issues early in the design process, such as ensuring the circuit can tolerate different types of faults and minimizing the impact of aging effects.

The next generation EDA flow aims to provide a more efficient, automated, and integrated design process that can meet the increasing demand for complex and high-performance integrated circuits in various applications. Next generation SoC architectures requires next generations EDA tools,

The integration of Large Language Models (LLMs) into Electronic Design Automation (EDA) presents both promising opportunities and significant challenges. While LLMs, such as GPT-4, have demonstrated remarkable capabilities in understanding and generating human-like text, their application to EDA tasks holds the potential for transformative solutions in design automation, error detection, and process optimization. However, the complexity of EDA tasks and the specialized nature of electronic design languages pose significant hurdles to the seamless integration of LLMs. Additionally, ethical considerations regarding data privacy, model reliability, and the potential for automation-induced obsolescence in skilled professions must be addressed. Current advancements in applying LLMs to EDA include various applications such as chip design optimization, Verilog code generation and debugging, and enhancing the design flow through natural language interfaces. For instance, frameworks like ChipNeMo aim to optimize LLM utilization in chip design by employing domain adaptation techniques, while tools like RTLFixer automate syntax-related error resolution in Verilog code using LLMs. These applications demonstrate the potential of LLMs to revolutionize different aspects of the EDA process. The future of LLM integration into EDA depends on several key factors. Firstly, the development of specialized LLMs tailored to understand electronic design languages and processes is crucial. This requires extensive domain-specific data and training to ensure that LLMs comprehend the specialized syntax and semantics of electronic design accurately. Secondly, addressing ethical and practical concerns surrounding data security, privacy, and responsible AI usage is essential to ensure the responsible integration of LLMs into EDA workflows. Looking ahead, there are promising avenues for further exploration, including the development of AI-assisted design tools, educational applications of LLMs in training electronic designers, and collaborative research between AI experts and EDA professionals. Such collaboration can accelerate the development of customized LLMs that cater specifically to the needs of the EDA industry and open new frontiers in AI application in electronics [31]. Python plays a significant role in Deep learning [38].

4.2 Deep Learning for Front-End Design

The typical front-end design flow starts with design specification, followed by HDL coding, verification through techniques like simulation and formal methods, logic synthesis to transform HDL to gate-level netlists, and logic optimization while performing equivalence checking. The development of front-end EDA tools involving HDLs, high-level synthesis, advanced verification methodologies, and logic synthesis has revolutionized digital design by automating tedious manual tasks, enabling more complex designs, optimizing for power/performance/area, and significantly mproving design productivity compared to the era of hand-drawn schematics [41].

4.2.1 Deep Learning for System Specification and Pre-RTL/ Architecture Design Exploration

Deep Learning (DL) has emerged as a promising tool for system-on-chip (SoC) architecture design exploration. DL techniques can enable the optimization of SoC architectures, improve their performance, reduce their power consumption, and accelerate their time-to-market. Here are some ways DL can be used for SoC architecture design exploration:

- **Automated Design Space Exploration (DSE)**: DL techniques can be used to automate DSE by reducing the design space and improving the search efficiency. For instance, Reinforcement Learning (RL) can be used to optimize the design parameters of an SoC, such as its memory hierarchy or its interconnect topology.
- **Hardware/Software Co-Design**: DL techniques can be used to optimize the hardware/ software co-design of a SoC. For example, DL models can be trained to predict the performance of different hardware/software configurations and guide the design process accordingly.
- **Power Optimization**: DL techniques can be used to optimize power consumption by modeling the power consumption of different components in a SoC and identifying the

most power-hungry components. For example, Convolutional Neural Networks (CNNs) can be used to analyze power traces and identify power-hungry components.

- **Performance Optimization**: DL techniques can be used to optimize the performance of an SoC by predicting the performance of different architectures and guiding the design process accordingly. For example, GANs can be used to generate synthetic workloads that can be used to test and optimize SoC designs.
- **Area Optimization**: DL can be exploited to optimize area.
- **Fault Detection and Diagnosis**: DL techniques can be used to detect and diagnose faults in a SoC. For example, Recurrent Neural Networks (RNNs) can be used to analyze system-level data such as sensor readings and user interactions and identify anomalous behavior that may indicate a fault.

The authors in [36] introduce an end-To-end approach for implementing a chatbot system that helps developers to find the information they need from protocol specifications (pdf documents). The proposed chatbot is a retrieval-based model which uses machine learning algorithms in predicting answers based on the provided training-set. The proposed system is based on the RASA framework and cdQA library using BERT model. We discuss the pre-processing phase for our dataset, the implementation phase, and how we fine-tune the model to have better performance mentioning all the used tools and the methodology our work is based on. The proposed chatbot can get users questions in voice and text forms and return answers in both forms too. The user should provide the protocol name as a keyword, so the chatbot can retrieve the answer, otherwise the chatbot asks the user to provide the protocol name. The response time of the chatbot is very reasonable. The model performance was enhanced after deploying fine–tuning.

4.2.2   Deep Learning for Logic/RTL Design

Deep Learning has shown promising results in improving the efficiency and automation of logic/RTL design, including tasks such as RTL generation, **optimization**, and verification. By using deep learning models, it is possible to learn complex patterns and correlations in the design space, which can be used to guide the design process and generate optimized designs. For example, deep learning can be used to generate RTL code from high-level descriptions such as C/C++ or MATLAB, reducing the time and effort required for manual RTL coding. Deep learning can also be used for optimizing RTL designs by predicting the best configuration for various design parameters such as clock frequency, power consumption, and area. Additionally, deep learning can be used for RTL verification by predicting potential design errors or mismatches. Moreover, DL can accelerate the RTL design by filling the gap in incomplete code [18, 19].

Authors in [22] introduce a novel approach to improving EDA tools through Verilog code **autocompletion** using deep learning techniques. It addresses the repetitive and time-consuming nature of writing Verilog code by proposing a deep learning framework for training a Verilog autocompletion model. The framework involves pretraining models on general programming language data and fine-tuning them on a Verilog dataset curated from open-source repositories. Key points include the identification of challenges in Verilog code writing, the creation methodology of the Verilog dataset, the proposed deep learning framework involving pretraining and fine-tuning, experimental validation demonstrating improved autocompletion accuracy, and distinctions from previous studies [23].

The paper in [28] introduces RTLFixer, a framework designed to **automatically fix syntax errors** in Verilog code generated by Large Language Models (LLMs). It addresses the issue of approximately 55% of errors in LLM-generated Verilog being syntax-related, which can cause compilation failures. RTLFixer employs Retrieval-Augmented Generation (RAG) and ReAct prompting, enabling LLMs to act autonomously in debugging code with feedback. Through this framework, about 98.5% of compilation errors in a debugging dataset are successfully corrected, leading to significant improvements in pass rates in benchmark evaluations. The integration of human expertise stored in a retrieval database enhances error correction through RAG, and the approach demonstrates promising results in addressing syntax errors, thereby streamlining the Verilog coding process for both LLMs and human engineers.

4.2.3   Deep Learning for High Level Synthesis

One of the objectives of ML algorithms in HLS is to estimate the performance results in terms of resource and time optimization. The time optimization problem can be modelled as a regression problem, and then used the deep learning training to evaluate the clock frequency of the HLS tool's output [10]. Machine learning in HLS can be used to explore the design space too [11]. Deep learning is revolutionizing the way digital systems are designed and synthesized. High-level synthesis (HLS) is a crucial step in this process, as it transforms a behavioral description of a system into a Register Transfer Level (RTL) implementation. Deep learning techniques have been used to improve the efficiency and accuracy of this process. Deep Learning for High-Level Synthesis refers to the use of deep learning techniques in automating the design process of digital circuits. High-Level Synthesis (HLS) is a design methodology that allows designers to describe the desired behavior of a digital circuit at a high level of abstraction and then automatically synthesize the circuit from the high-level description. Deep learning techniques have been applied to HLS to address challenges such as optimizing the design for power and performance, reducing the design time and complexity, and improving the quality of the design. There are several ways in which deep learning can be applied to HLS:

1. Performance prediction: Deep learning can be used to predict the performance of an RTL implementation based on the behavioral description. This can help designers make informed decisions about the design parameters.
2. Architecture exploration: Deep learning can be used to explore different architectures and configurations for a given system. This can help designers find the most efficient and optimal design.
3. Code optimization: Deep learning can be used to optimize the RTL implementation by identifying the critical path and suggesting modifications to the code.
4. Verification: Deep learning can be used to verify the RTL implementation by comparing it to the behavioral description and identifying any discrepancies.

4.2.4 Deep Learning for Functional Verification

Machine learning can be used to **generate new tests** to cover desired cover points that are considered hard to cover with manually written tests [9]. Supervised and reinforcement ML algorithms to guide the stimulus generator to hit planned coverage metrics. Functional verification is a critical step in the design process of complex system-on-chips (SoCs), ensuring that the design meets the specified functionality and performance requirements. Deep learning has shown potential in improving the efficiency and effectiveness of functional verification by automating certain aspects of the verification process. One application of deep learning in functional verification is the use of artificial neural networks (ANNs) to predict the expected behavior of a design. ANNs can be trained on a set of input–output pairs, representing correct behavior of the design, and then used to predict the expected output for new inputs. This can help identify design bugs or errors in the verification process. Another application is the use of deep reinforcement learning (DRL) to optimize the verification process. DRL agents can be trained to automatically generate test cases and perform dynamic verification based on feedback from the design. This can help reduce the time and effort required for manual verification, while also improving coverage of the design space. DL can be used to speedup coverage closure [37].

Utilizing the capabilities of Machine Learning and Artificial Intelligence has become increasingly imperative in various domains. A prime example lies in the realm of Automated Test Generation, where the burgeoning complexity of contemporary hardware devices poses significant challenges. Traditional methods like constrained random stimulus generation have proven insufficient in fully simulating the intricate functionalities of digital designs. However, the advent of Reinforcement Learning heralds a paradigm shift, offering a novel means to efficiently leverage computational resources. By delving into the nuanced interplay between configuration parameters, stimuli applied to digital design inputs, and their resulting functional states, Reinforcement Learning uncovers subtle correlations that were previously elusive. Deploying ML to generate new tests needs much fewer steps to attain a target state compared to conventional constrained random stimulus generation methods. This underscores the profound impact and potential of AI-driven techniques in revolutionizing testing methodologies and enhancing the robustness of digital designs in today's increasingly complex technological landscape [20, 21].

The authors in [24] address the challenges in verifying VLSI chip designs, emphasizing the significant time and resources required for manual verification, which can consume up to 70% of the design effort. To enhance efficiency and reliability, the authors propose AUTG (Automatic UVM Testbench Generator), **a tool aimed at automating the generation of Universal Verification Methodology** (UVM)-based Testbenches. AUTG takes a golden reference (GRef) file and the port list of the Design under Test (DUT) as inputs to generate a complete UVM-based Testbench, including reference model transformation and checkers creation. Unlike existing tools, AUTG prioritizes usability, cost-effectiveness, and feature-richness. Implemented and tested with various benchmarks, AUTG demonstrates its ability to successfully generate functional verification code, effectively hunting bugs in defective RTL designs [39].

The authors in [25] presents AAG (Automatic Assertion Generation), a framework designed to address the challenges in functional verification of modern hardware designs, particularly due to their increasing complexity and size. AAG aims to **automate the process of generating assertions** for Register Transfer Level (RTL) code written in Verilog, thereby simplifying verification tasks for engineers. Traditional simulation-based verification methods are time-consuming and prone to missing corner cases, leading to potential bugs in the design. Formal verification methods have limitations in expressing specifications and handling large designs. Assertion Based Verification (ABV) emerges as a promising approach, using temporal language expressions called assertions for specification. However, manual assertion writing poses challenges, especially for complex specifications. AAG proposes to automate assertion generation using GoldMine, leveraging data mining algorithms to generate assertions from simulated data. The framework includes a user-friendly GUI to facilitate testbench creation and simulation, enhancing the quality of assertion generation for ABV [29, 32].

The authors in [40] introduce COVERUP, a novel system designed to enhance the generation of high-coverage Python regression tests by integrating **coverage analysis** with large-language models (LLMs). COVERUP iteratively refines test suites by analyzing coverage and engaging in dialogues with LLMs to focus on uncovered code segments. Compared to existing systems like CODAMOSA, COVERUP significantly improves coverage metrics, achieving median line coverage of 81%, branch coverage of 53%, and line + branch coverage of 78%. By leveraging detailed coverage information and iterative refinement, COVERUP demonstrates substantial improvements in test coverage, positioning itself as a promising advancement in automated test generation for Python programs.

4.2.5   Deep Learning for Logic Synthesis: Synthesized Logic/Netlist Optimization and Logic Size/Depth Reduction

Logic synthesis is the step where the logic needs to be translated into logic primitives such as NAND, NOR, XOR, and INV. Netlists are used to represent the implementation of logic circuits. Deep learning can be automatically deployed to algebraic logic optimization algorithms used in synthesis EDA tools. For example, it can find optimum representations for n-input functions represented in sum-of-products (SOP) format. using graph convolutional network and deep reinforcement learning with a reward function which is defined with respect to the optimization objective [1]. Moreover, it can minimize the gate count of a logic circuit at synthesis level. DL algorithms will work on a graph representation of logic circuits [2]. Optimization can be applied to logic size which impacts the area improvement or logic depth which decreases the hardware's latency for increased speed. Logic synthesis sequence can be formulated using Markov decision process (**MDP**) or Markov Chain Monte Carlo optimization [16].

The utilization of Reinforcement Learning (RL) in the logic synthesis problem arises due to the need for optimizing Boolean logic while preserving its functionality. Logic synthesis involves representing the logic as a logic graph, such as an and-inverter graph (AIG), and then performing graph operations to optimize the size and depth of the graph without altering the logic's function. In practice, a sequence of different graph operations is applied to the logic graph to find solutions. However, determining the best sequence of optimization steps is often not thoroughly explored. These graph operations involve replacing sub-graphs with equivalent logic representations.

The result of each operation is determined based on the current structure of the graph. Consequently, the logic optimization process can be seen as a dynamic decision-making process. To apply RL to the logic synthesis problem, we formulate it as a Markov Decision Process (MDP). Given an initial logic graph, we aim to make a certain number of sub-graph optimization steps, and our objective is to achieve the desired outcome after these steps. At each step, we observe the current logic graph and select a graph operation from a set of candidate actions.

This formulation aligns well with the typical framework of an MDP. By employing RL techniques, we can train an RL agent to learn a policy that selects the most appropriate graph operation at each step based on the observed graph state. The agent receives rewards based on the quality of the resulting graph after the sequence of operations. The RL agent then learns to maximize these rewards over time, leading to improved logic synthesis outcomes. Applying RL to the logic synthesis problem allows for the exploration of different optimization strategies and the discovery of novel sequence patterns that may lead to better results. RL-based approaches have the potential to automate and enhance the process of logic optimization, aiding in the development of more efficient and compact logic circuits [17].

4.2.6   Deep Learning for Static Timing Analysis/Early RTL Delay Prediction

Several deep learning techniques can be used to model cell delays thus predicting the static timing analysis [6]. Input/output voltage waveforms can be used as training data. Early estimations of the delay information of a design at RTL can ease the design-space exploration to find the optimal implementation and architectural option [8]. STA is a critical step in digital circuit design, which ensures that the timing constraints of a design are met. Deep learning can help in reducing the computational time required for STA by predicting the slack values of a design. One approach is to use a convolutional neural network (CNN) to learn the relationship between the input features and the slack values. The input features can include timing constraints, circuit topology, and layout information. The CNN is trained on a large dataset of designs to predict the slack values for new designs. Another approach is to use a recurrent neural network (RNN) to model the timing paths in a design. The RNN can predict the slack values for each path and combine them to obtain the overall slack value. This approach can handle designs with complex timing paths and can be trained using reinforcement learning. Deep learning for STA is still a relatively new area of research, and there is ongoing work to improve the accuracy and efficiency of the models.

4.2.7   Deep Learning for Formal Verification

ML can enhance traditional FV approaches such as static analysis (SA), model-checking, theorem-proving (TP), and SAT solving. In SAT solving, ML models are employed to predict runtime, select restart strategies, choose branching variables, determine the best solving algorithms, and configure solver parameters. These tasks involve both classification and regression ML approaches to optimize solver performance and efficiency. Similarly, in TP, ML techniques aid in selecting relevant facts, configuring proof engine parameters, guiding interactive proofs, enhancing automation and efficiency in theorem proving tasks. Furthermore, in SA, ML is utilized to identify actionable alerts by classifying alarms and predicting bugs from previous code versions, thereby assisting developers in prioritizing code reviews and allocating resources effectively [43].

4.3   Deep Learning for Back-End Design

Back-End Design involves transitioning from a gate-level netlist to the final layout. It begins with technology mapping to map the design to a specific process library. The physical design stage includes floorplanning to determine the chip's physical layout, power delivery network design, placement of standard cells/IP blocks, clock tree synthesis for synchronized timing, and routing to connect components. As chip complexity increases, additional considerations like thermal analysis, design for yield strategies to maximize manufacturing robustness, and physical verification become crucial. Physical verification comprises design rule checking for manufacturability, electrical rule checking for functional correctness, and layout vs schematic verification to ensure the layout matches the original circuit design. The back-end flow optimizes for performance, power, area, signal integrity and prepares the design for manufacturing through methodical layout implementation and stringent verification checks.

4.3.1 Deep Learning for Routing Congestion Prediction at Floor Planning or Placement Stage

Congestion occurs when demand for the routing resources is high in certain regions. Congestion can cause routing failure. ML algorithms can be deployed to calculate and predict routing congestion for a given VLSI design. Then congestion mitigation techniques are deployed to alleviate the demand of routing resources in these areas. Supervised ML such as regression or fully convolutional network approaches can be adapted to estimate congestion by mapping netlist and interconnect grid features onto pixelated images and regressing on them given a target image label [5]. Routing congestion prediction can be done at Floor Planning or Placement stage to reduce the total design time cost as Congestion prediction model can guides the placement or the floor-planning. The congestion metrics are used for the training data as labels [7, 12, 13].

4.3.2 Deep Learning for Power Estimation/Prediction

Power estimation is very important step as the IR-drop might create time infringements [4]. The netlist is used for feature extraction. Regression technique of ML can be used to estimate the power consumption of the MOSFET-based digital circuits [3]. Power estimation is an important step in design optimization and power management. Accurate power estimation is essential for reducing power consumption and improving battery life in portable devices. One approach to using deep learning for power estimation is to use a convolutional neural network (CNN) to learn the relationship between the input features and the power consumption of a design. The input features can include circuit topology, layout information, voltage and current waveforms. The CNN can be trained on a large dataset of designs to predict the power consumption for new designs. Another approach is to use a recurrent neural network (RNN) to model the power consumption over time. The RNN can predict the power consumption for each cycle and combine them to obtain the overall power consumption. This approach can handle designs with dynamic power consumption and can be trained using reinforcement learning.

The paper in [26] introduces EDALearn, a comprehensive open-source benchmark suite designed for Machine Learning (ML) tasks in Electronic Design Automation (EDA) for Very Large-Scale Integration (VLSI) design. Addressing the lack of extensive public datasets for effective ML models in EDA, EDALearn offers a holistic flow from synthesis to physical implementation, enriching data collection across various stages of the design process. It accommodates a wide range of VLSI design instances and sizes, providing a representative complexity of modern VLSI designs. The suite includes critical EDA tasks such as placement, routing, power analysis, timing analysis, and IR drop prediction, along with a standardized evaluation framework for fair assessment of ML models. EDALearn aims to foster reproducibility, promote research into ML transferability across different technology nodes, and encourage further advances in the ML-EDA domain by offering in-depth data analysis and comprehensive coverage of design instances. The benchmark suite's contributions are outlined, including its end-to-end flow, open-source reference flow, diverse design instances, and thorough data analysis, aiming to facilitate collaboration, knowledge sharing, and advancement in ML-based approaches for EDA.

EDALearn benchmark suite encompasses various critical Electronic Design Automation (EDA) tasks, including power prediction, slack time prediction, routability prediction, and IR drop prediction [27].

- **Power prediction**: Estimating circuit power consumption, vital for preventing overheating and ensuring reliability, particularly in mobile devices.
- **Slack time prediction**: Determining task scheduling flexibility without delaying the project, aiding in identifying potential timing violations.
- **Routability prediction**: Assessing the feasibility of routing a design within specified constraints, minimizing iterative refinements.
- **IR drop prediction**: Estimating voltage drop across the power distribution network, crucial for performance and reliability, enabling informed power optimization decisions.

4.3.3 Deep Learning for Physical Design and Verification

There are various categories of AI/ML applications in physical design, including prediction, optimization, and generation, each with its specific challenges and successes. For instance, prediction involves supervised learning to predict design quality-of-results metrics like power consumption, timing, and congestion, while optimization applies Bayesian optimization or reinforcement learning directly to EDA problems. Generation utilizes generative models like GANs or transformers to produce solutions to design problems directly. However, generative approaches face difficulties in consistently delivering well-formed solutions, often requiring enhancement with traditional methods [30]. The authors in [44] propose a novel methodology leveraging Large Language Models (LLMs) to enhance the optimization of standard cell layouts, crucial for modern digital circuit designs advancing towards 2 nm technology nodes. The process involves using the natural language and reasoning capabilities of LLMs to incrementally generate high-quality clustering constraints, optimizing the Performance-Power-Area (PPA) and improving routability of the cell layouts. By incorporating the expertise of human designers and employing ReAct prompting techniques, the method addresses the complexity and design rule constraints inherent in advanced nodes.

4.4 Deep Learning for Fabrication

The proposed framework in [33] presents a comprehensive Artificial Intelligence (AI) and Deep Learning (DL)-based image analysis approach for hardware assurance of digital integrated circuits (ICs), aiming to **examine/inspect** and verify various hardware information from **Scanning Electron Microscope (SEM) images**. Unlike previous methods that relied on manual analysis or classical image processing techniques, the proposed framework leverages DL-based methods at all essential steps of the analysis. This approach marks a significant departure from existing practices and demonstrates several novel features. Firstly, the framework introduces innovative DL-based methods for tasks such as stitching misalignment detection and stacking movement regression, which were previously performed manually or with classical techniques. These methods offer advantages in terms of speed, accuracy, and robustness against noise, thereby significantly enhancing the automation level in hardware assurance tasks. Additionally, the framework emphasizes the use of automated and semi-automated methods for preparing training data, including the utilization of synthetic data to train DL models. This approach addresses the challenge of obtaining high-quality and large quantities of training data, which is crucial for the effectiveness of DL-based methods. Furthermore, the framework proposes various DL model architectures tailored to specific image analysis tasks, ensuring that the models can operate on feature images without the need for retraining. This flexibility maximizes model reusability and facilitates the integration of DL-based methods into hardware assurance workflows. By applying the proposed framework to analyze SEM images of a large digital IC, the efficacy of DL-based methods in automating analysis tasks is demonstrated, highlighting their speed, accuracy, and robustness.

To **detect hardware trojan** in layout, SEM images of manufactured ICs need to be compared with their original GDS images. Authors in [34] propose a deep learning framework to detect malicious modifications In IC Layout. In the proposed framework, authors convert Scanning Electron Microscope (SEM) images to Graphics Data System (GDS) images for layout modification detection. Leveraging our customized DL segmentation model, a two-model transformation strategy, and a tailored loss function, SEM2GDS accurately transforms SEM images into GDS format, ensuring sharp-cornered shapes for direct comparison with original GDS images.

4.5   Deep Learning for PCB

The work in [42] highlights the necessity for specialized artificial intelligence (AI) techniques tailored for **counterfeit and defect detection** of Printed Circuit Board (PCB) components, emphasizing the limitations of popular computer vision object detection methods in scenarios requiring high accuracy amidst dense, low inter-class/high intra-class variation, and limited-data hardware assurance contexts. The authors introduce the Electronic Component Localization and Detection Network (ECLAD-Net) as a proposed solution, comparing its performance with existing methodologies such as region-based convolutional neural networks (RCNNs) and single-shot detectors (SSDs). Moreover, **AI-based PCB schematic** verification is a promising approach to improve the accuracy, efficiency, and reliability of the design process. By leveraging machine learning and artificial intelligence, this technology can automatically detect errors, suggest corrections, and ensure adherence to design rules and standards.

4.6   The Future of EDA Tools: Digital Twin

The future of simulation in engineering and various other domains isn't solely dependent on Artificial Intelligence (AI), but it also involves the concept of digital twins. Digital twins represent a paradigm shift in how we model, understand, and interact with complex systems, offering tremendous potential for predictive maintenance and real-time insights. A digital twin is a virtual representation of a physical asset, system, or process that mirrors its real-world counterpart. It combines real-time data from sensors, IoT devices, and other sources with advanced modeling and simulation techniques to create a dynamic, digital replica of the physical system. Digital twins provide a holistic view of the asset's behavior, performance, and condition throughout its lifecycle, enabling better decision-making, optimization, and predictive analysis. By capturing data on how assets operate in real-time, digital twins facilitate proactive maintenance, troubleshooting, and optimization strategies, ultimately improving reliability, uptime, and operational efficiency. Summary of its applications [35]:

- **Predictive Maintenance**
  - One of the key applications of digital twins is predictive maintenance, where AI algorithms analyze real-time data from the digital twin to anticipate potential equipment failures or performance degradation.

- By monitoring factors such as temperature, vibration, pressure, and usage patterns, predictive maintenance algorithms can identify early warning signs of impending issues and recommend timely interventions to prevent unplanned downtime and costly repairs.
  - Predictive maintenance strategies based on digital twins help organizations transition from reactive and scheduled maintenance approaches to more proactive and condition-based methodologies, saving time, resources, and operational costs in the long run.
- **Real-Time Insights**
  - Digital twins provide engineers, operators, and stakeholders with real-time insights into the behavior and performance of assets, systems, or processes.
  - Through visualization dashboards, analytics tools, and predictive models, users can monitor key performance indicators, analyze trends, and respond promptly to changing conditions or anomalies.
  - Real-time insights derived from digital twins enable continuous optimization, adaptive control, and data-driven decision-making, empowering organizations to maximize productivity, efficiency, and competitiveness in dynamic environments.

4.7   Conclusions

Machine learning and deep learning provides huge potential for improving existing chip design and verification. Machine learning (ML) is a rapidly growing field that has found applications in numerous industries, including electronic design automation (EDA). One of the recent developments in this area is ML-assisted simulators that can improve the accuracy and efficiency of simulation tools. In this chapter, we discussed various aspects of ML-assisted simulators and their potential applications in EDA. While there are challenges associated with implementing these tools, there is no doubt that they will play an increasingly important role in the future of chip design. As the amount of data generated by EDA tools continues to grow, ML-assisted simulators will become an essential tool for designers to ensure that their designs meet the performance and power targets.

References

1. Haaswijk, W., et al. (2018). Deep learning for logic optimization algorithms. In *2018 IEEE International Symposium on Circuits and Systems (ISCAS)*, Florence, Italy, 2018 (pp. 1–4). https://doi.org/10.1109/ISCAS.2018.8351885
2. Timoneda, X., & Cavigelli, L. (2021). Late breaking results: Reinforcement learning for scalable logic optimization with graph neural networks. In *2021 58th ACM/IEEE Design Automation Conference (DAC)*, San Francisco, CA, USA, 2021 (pp. 1378–1379). https://doi.org/10.1109/DAC18074.2021.9586206

3. Bhavesh, M. D., Anilkumar, N. A., Patel, M. I., Gajjar, R., & Panchal, D. (2022). Power consumption prediction of digital circuits using machine learning. In *2022 2nd International Conference on Artificial Intelligence and Signal Processing (AISP)*, Vijayawada, India, 2022 (pp. 1–6). https://doi.org/10.1109/AISP53593.2022.9760542

4. Poovannan, E., & Karthik, S. (2023). Power prediction of VLSI circuits using machine learning. *CMC—Computers Materials & Continua, 74*(1), 2161–2177.

5. Zhou, Z., Zhu, Z., Chen, J., Ma, Y., Yu, B., Ho, T., Lemieux, G., & Ivanov, A. (2019). Congestion-aware global routing using deep convolutional generative adversarial networks. In *2019 ACM/IEEE 1st Workshop on Machine Learning for CAD (MLCAD)*, 2019 (pp. 1–6).

6. Bian, S., Hiromoto, M., Shintani, M., & Sato, T. (2017). LSTA: Learning-based static timing analysis for high-dimensional correlated on-chip variations. In *2017 54th ACM/EDAC/IEEE Design Automation Conference (DAC)*, Austin, TX, USA, 2017 (pp. 1–6). https://doi.org/10.1145/3061639.3062280

7. Zhao, J., Liang, T., Sinha, S., & Zhang, W. (2019). Machine learning based routing congestion prediction in FPGA high-level synthesis. In *2019 Design, Automation & Test in Europe Conference & Exhibition (DATE)*, Florence, Italy, 2019 (pp. 1130–1135). https://doi.org/10.23919/DATE.2019.8714724

8. Lopera, D. S., Servais, L., Prebeck, S., & Ecker, W. (2022). Early RTL delay prediction using neural networks. *Microprocessors and Microsystems, 94*.

9. Vasudevan, S., Jiang, W. J., Bieber, D., Singh, R., Richard Ho, C., Sutton, C. (2021). Learning semantic representations to verify hardware designs. *Advances in Neural Information Processing Systems, 34*, 23491–23504.

10. Makrani, H. M., Farahmand, F., Sayadi, H., Bondi, S., Dinakarrao, S. M. P., Homayoun, H., & Rafatirad, S. (2019). Pyramid: Machine learning framework to estimate the optimal timing and resource usage of a high-level synthesis design. In *Proceedings of the 2019 29th International Conference on Field Programmable Logic and Applications (FPL)*, Barcelona, Spain, September 8–12, 2019 (pp. 397–403).

11. Goswami, P., Schaefer, B. C., & Bhatia, D. (2023). Machine learning based fast and accurate High Level Synthesis design space exploration: From graph to synthesis. *Integration, 88*, 116–124.

12. Al-Hyari, A., Szentimrey, H., Shamli, A., Martin, T., Grewal, G., & Areibi, S. (2021). A deep learning framework to predict routability for FPGA circuit placement. *M Transactions on Reconfigurable Technology and Systems (TRETS), 14*, 1–28.

13. He, Y., Li, H., Jin, T., & Bao, F. S. (2022). Circuit routing using Monte Carlo tree search and deep reinforcement learning. In *Proceedings of the 2022 International Symposium on VLSI Design, Automation and Test (VLSI-DAT)*, Hsinchu, Taiwan, April 18–21, 2022 (pp. 1–5).

14. Mohamed, K. S. (2018). Brain-inspired machine learning algorithm: Neural network optimization. In *Machine learning for model order reduction*. Springer. https://doi.org/10.1007/978-3-319-75714-8_6

15. Mohamed, K. S. (2020). *Neuromorphic computing and beyond*. Springer.

16. Budak, A. F., et al. (2022). Reinforcement learning for electronic design automation: Case studies and perspectives. In *2022 27th Asia and South Pacific Design Automation Conference (ASP-DAC)*. IEEE.

17. Zhu, K., Liu, M., Chen, H., Zhao, Z., & Pan, D. Z. (2020). Exploring logic optimizations with reinforcement learning and graph convolutional network. In *2020 ACM/IEEE Workshop on Machine Learning for CAD (MLCAD)*.

18. Pearce, H., Tan, B., & Karri, R. (2020). DAVE: Deriving automatically Verilog from English. In *Proceedings of the 2020 ACM/IEEE Workshop on Machine Learning for CAD*. ACM, November 2020 (pp. 27–32) [Online]. Available at: https://doi.org/10.1145/3380446.3430634

19. Thakur, S. et al. (2022). Benchmarking large language models for automated Verilog RTL code generation. In *2023 Design, Automation & Test in Europe Conference & Exhibition (DATE)* (pp. 1–6).
20. Mohamed, K. S. (2023). *Deep learning-powered technologies: Autonomous driving, artificial intelligence of things (AIoT), augmented reality, 5G communications and beyond.* Springer.
21. Dinu, A., & Ogrutan, P. L. (2022). Reinforcement learning made affordable for hardware verification engineers. *Micromachines, 13*(11), 1887.
22. Dehaerne, E., et al. (2023). A deep learning framework for Verilog autocompletion towards design and verification automation. arXiv preprint arXiv:2304.13840
23. Pei, Z., et al. (2024). BetterV: Controlled Verilog generation with discriminative guidance. arXiv preprint arXiv:2402.03375
24. Ismael, M., Hroub, A., & Abu-Issa, A. (2023). AUTG: An automatic UVM-based TestBench generator for VLSI chip design verification. In *2023 International Conference on Microelectronics (ICM)*. IEEE.
25. Murtza, S. A., Hasan, O., & Saghar, K. (2018). Aag: An automatic assertion generation framework for RTL designs. In *2018 International Conference on Computing Mathematics and Engineering Technologies (iCoMET)* (pp. 1–6).
26. Pan, J., et al. (2023). EDALearn: A comprehensive RTL-to-signoff EDA benchmark for democratized and reproducible ML for EDA research. arXiv preprint arXiv:2312.01674
27. Huang, G., Hu, J., He, Y., Liu, J., Ma, M., Shen, Z., Wu, J., Xu, Y., Zhang, H., Zhong, K., et al. (2021). Machine learning for electronic design automation: A survey. *ACM Transactions on Design Automation of Electronic Systems (TODAES), 26*(5), 1–46.
28. Tsai, Y. D., Liu, M., & Ren, H. (2023). Rtlfixer: Automatically fixing rtl syntax errors with large language models. arXiv preprint arXiv:2311.16543
29. Fang, W., et al. (2024). AssertLLM: Generating and evaluating hardware verification assertions from design specifications via multi-LLMs. arXiv preprint arXiv:2402.00386
30. Kahng, A. B. (2024). Solvers, engines, tools and flows: The next wave for AI/ML in physical design. In *Proceedings of the 2024 International Symposium on Physical Design*.
31. He, Z., & Yu, B. (2024). Large language models for EDA: Future or mirage?
32. Kande, R., Pearce, H., Tan, B., Dolan-Gavitt, B., Thakur, S., Karri, R., & Rajendran, J. (2023). LLM-assisted generation of hardware assertions. arXiv preprint arXiv:2306.14027
33. Lin, T., et al. (2021). Deep learning-based image analysis framework for hardware assurance of digital integrated circuits. *Microelectronics Reliability, 123*, 114196.
34. Lin, T., Shi, Y., & Gwee, B. H. (2023). SEM2GDS: A deep-learning based framework to detect malicious modifications in IC layout. In *2023 IEEE International Symposium on Circuits and Systems (ISCAS)*. IEEE.
35. Boschert, S., Heinrich, C., & Rosen, R. (2018). Next generation digital twin. In *Proceedings— TMCE* (Vol. 2018). Las Palmas de Gran Canaria.
36. ElSayed, H. S., Hussien, M. T., Salah, E. M., Farouk, A. K., Ahmed, A. T., Abd El-Hafez, M. M., & Salah K. (2023). Chatbot as a virtual assistant to retrieve information from datasheets using memory controllers domain knowledge. In *2023 30th IEEE International Conference on Electronics, Circuits and Systems (ICECS)* (pp. 1–7). IEEE.
37. Mohamed, K. S. (2024). Pre-silicon verification and post-silicon validation methodologies. *Heterogeneous SoC design and verification: HW/SW co-exploration, co-design, co-verification and co-debugging.* Springer (pp. 85–131).
38. Mohamed, K. S. (2023). Python for deep learning: A general introduction. In *Deep Learning-Powered Technologies: Autonomous Driving, Artificial Intelligence of Things (AIoT), Augmented Reality, 5G Communications and Beyond* (pp. 171–201). Springer.

39. Mahmoud, K., Ahmed, R., Ayman, K., Ayman, M., Taie, W., Ibrahim, Y., Mostafa, H., & Salah, K. (2022). Towards a generic UVM. In *2022 IEEE High Performance Extreme Computing Conference (HPEC)* (pp. 1–6). IEEE.

40. Pizzorno, J. A., & Berger, E. D. (2024). CoverUp: Coverage-guided LLM-based test generation. arXiv preprint arXiv:2403.16218

41. Chen, L., Chen, Y., Chu, Z., Fang, W., Ho, T.-Y., Huang, Y., Khan, S., et al. (2024). The dawn of AI-native EDA: Promises and challenges of large circuit models. arXiv preprint arXiv:2403.07257

42. Sathiaseelan, M., Azhagan, M., Paradis, O. P., Taheri, S., & Asadizanjani, N. (2021). Why is deep learning challenging for Printed Circuit Board (PCB) component recognition and how can we address it? *Cryptography, 5*(1), 9. https://doi.org/10.3390/cryptography5010009

43. Amrani, M., Lucio, L., & Bibal, A. (2018). ML + FV = $\heartsuit$? A survey on the application of machine learning to formal verification. arXiv: Software Engineering.

44. Ho, C.-T., & Ren, H. (2024). Large language model (LLM) for standard cell layout design optimization. arXiv preprint arXiv:2406.06549